Tile Patterns: Symmetry-Based Replication
by Joe Sweeney

Published by JS Blume Publishing, LLC.

ISBN: 978-1-960741-01-1

Table of Contents

Tile Patterns:
Symmetry-Based Replication

by Joe Sweeney

Introduction

When I was in high school, one of my favorite pastimes was creating small, simple patterns that would form larger ones when replicated. I called them "tile patterns" because I imagined what it would be like to place a simple pattern on a floor tile and then tile a floor with it.

I created the patterns on graph paper. A few years ago, I transferred the patterns to bitmap images and expanded my library. Altogether, I came up with 60 different patterns. I'm sure I missed many. So, I designed a web page where you can play around with creating patterns:

http://astutedata.com/tiles

If you create one that I already have, the page will let you know. Otherwise, you can save it, and I'll include the pattern (with credit where credit is due) in the next edition of this book.

Symmetry and Asymmetry

Wiktionary defines symmetry as:

1. Exact correspondence on either side of a dividing line, plane, center, or axis.

In two-dimensional space, this means identifying the center point of an image and either rotating it around that point or drawing a line that divides the image horizontally, vertically, or diagonally.

Symmetry around a point means that rotating the image doesn't change its appearance. The symmetric order is a count of how many times a match occurs when rotated through a full circle. For example, there are four matches when rotating a square.

Symmetry about a line means that the line acts as a mirror – the half image on one side of the line is a reflection of the half image on the other side.

This book uses three symmetries:

1. Rotational symmetry about a point.
2. Reflection symmetry about a horizontal line.
3. Reflection symmetry about a vertical line.

The opposite of symmetry is asymmetry. An image is much more interesting if it is asymmetric on one or more axes because the tiled pattern changes when a reflection or rotation transforms it.

Interestingly enough, it turns out that if an image is:

- Asymmetric for all three, there is no difference between the tiled pattern with just rotational symmetry applied and the tiled pattern with both reflection symmetries applied.

- Asymmetric on both reflections and symmetric rotationally with a degree of four, the double reflection vertically and horizontally does not create a unique tile pattern.

Generating Tile Patterns

As I started this book, I realized that I hadn't fully considered all possible combinations of rotations and reflections. My original graph paper tiles only rotated the simple patterns.

Given three types of transformations (rotation, horizontal reflection, and vertical reflection):

- Simple patterns with no asymmetries would form only one tile pattern.
-
- Simple patterns with a horizontal or vertical asymmetry have two tile variations each – the original and the asymmetric one.
-
- Simple patterns with two asymmetries would form four tile variations: the original, one for each asymmetry, and one with both asymmetries applied. Additionally, there are three ways to pair up the tile variations.
-
- Simple patterns with three asymmetries would yield eight tile variations: the original, one for each asymmetry, one for each possible pairing of asymmetries, and one for all asymmetries.

On top of all that, the simple patterns with multiple variations could be combined and arranged in various ways. Preliminary calculations indicated that my 61 simple patterns would form hundreds of tile patterns. It would have taken me a very long time to create each one manually.

Since I am a computer programmer, I wrote a program that reads each bitmap one at a time and generates the tile patterns.

Even still, this was complicated, and I had to make some decisions for handling simple patterns with two or more asymmetries:

1. For simple patterns with three asymmetries, I would not produce a tile pattern with all three applied. It would most likely duplicate one of the other combinations.

2. For the horizontal and vertical reflection pairing, I created a square from four squares. I placed the unaltered simple pattern in the upper left corner, and filled the other squares by flipping right, flipping down, and flipping left.

3. For the rotational asymmetry, I created a square from four squares. I placed the unaltered simple pattern in the upper-left corner and filled the other squares by rotating around the center point of the larger square.

4. For simple patterns with three asymmetries, I found that the tile pattern with rotation was identical to that with both horizontal and vertical reflections.

As a result of these decisions, I've eliminated some obvious, simple ones to have an easy method for auto-generating the bitmap files.

The original bitmap files are 17 pixels by 17 pixels. The program does the following:

- Imports the files and converts the bits to strings, using 0 for black and 1 for white, and stores the strings in a database.

- Checks the strings for symmetry and assigns a three-letter code.

- Generates all possible four-square patterns for each combination of tile variations as described earlier, adding only the unique ones to the database.

- Creates a bitmap 15 times the size, with a border, for the left page.

- Creates a bitmap that is 10 times larger and patterned 6 across and 10 down, with a border, for the right page.

Conventions Used in this Book

This book contains two types of images.

The simple pattern image appears on a left-side page and has a seven-character name:

1. The letter “T” for tile.
2. An "S" or "A" indicates whether it has horizontal symmetry or asymmetry.
3. An "S" or "A" indicates whether it has vertical symmetry or asymmetry.
4. An "S" or "A" indicates whether it has rotational symmetry or asymmetry.

The last three characters are a three-digit, left-zero-filled sequence number.

The tiled pattern image appears on a right-hand page (opposite its simple pattern) and has a ten-character name. The first seven characters are the same as the simple pattern upon which it is based. The last three characters are:

8. An "S" if horizontal symmetry has been applied; otherwise, an "A".
9. An “S” if vertical symmetry has been applied; otherwise an “A”.
10. An "S" if rotational symmetry has been applied; otherwise, an "A".

Organization of this Book

The simple patterns are grouped in sections based on the number of asymmetries the pattern contains.

The first chapter consists of those simple patterns that exhibit none of the three symmetries: rotational, horizontal, or vertical. These simple patterns only form one tiled pattern.

The second chapter consists of simple patterns that exhibit a single asymmetry: rotational or reflectional.

The third chapter consists of simple patterns that exhibit two asymmetries: rotational and one or both reflections.

The fourth chapter consists of simple patterns that exhibit three asymmetries: rotational and both reflections. These simple patterns form the widest variety of tiled patterns.

Within each chapter, a given simple pattern is shown with all possible tiled patterns.

Chapter 1: No Asymmetries

The simple patterns in this chapter exhibit three symmetries: rotational, horizontal, or vertical. These simple patterns form only one tiled pattern because none of the transformations affect their appearance.

Tile Patterns: Symmetry-Based Replication

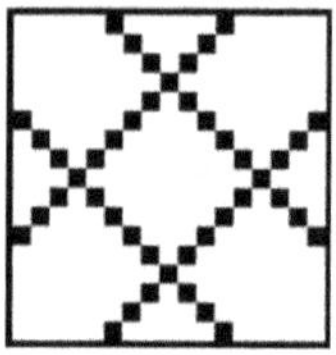

TSSS035

TSSS035SSS

Tile Patterns: Symmetry-Based Replication

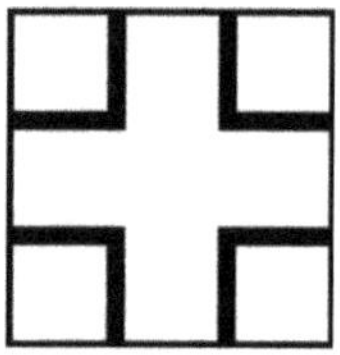

TSSS036

TSSS036SSS

Tile Patterns: Symmetry-Based Replication

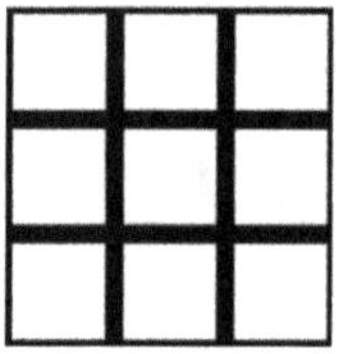

TSSS037

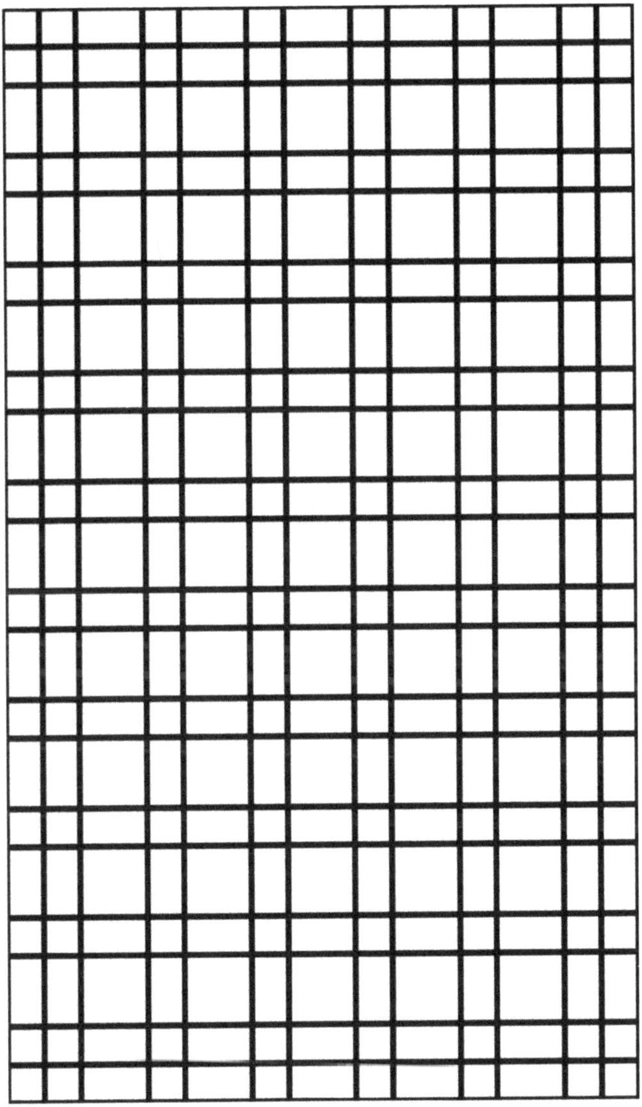

TSSS037SSS

Tile Patterns: Symmetry-Based Replication

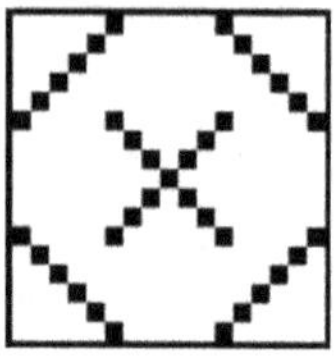

TSSS038

TSSS038SSS

Tile Patterns: Symmetry-Based Replication

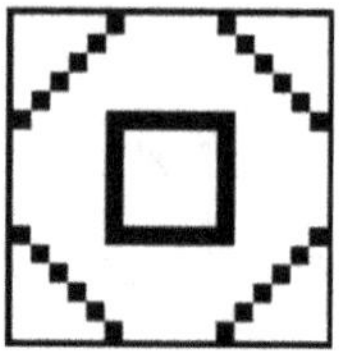

TSSS039

TSSS039SSS

Tile Patterns: Symmetry-Based Replication

TSSS040

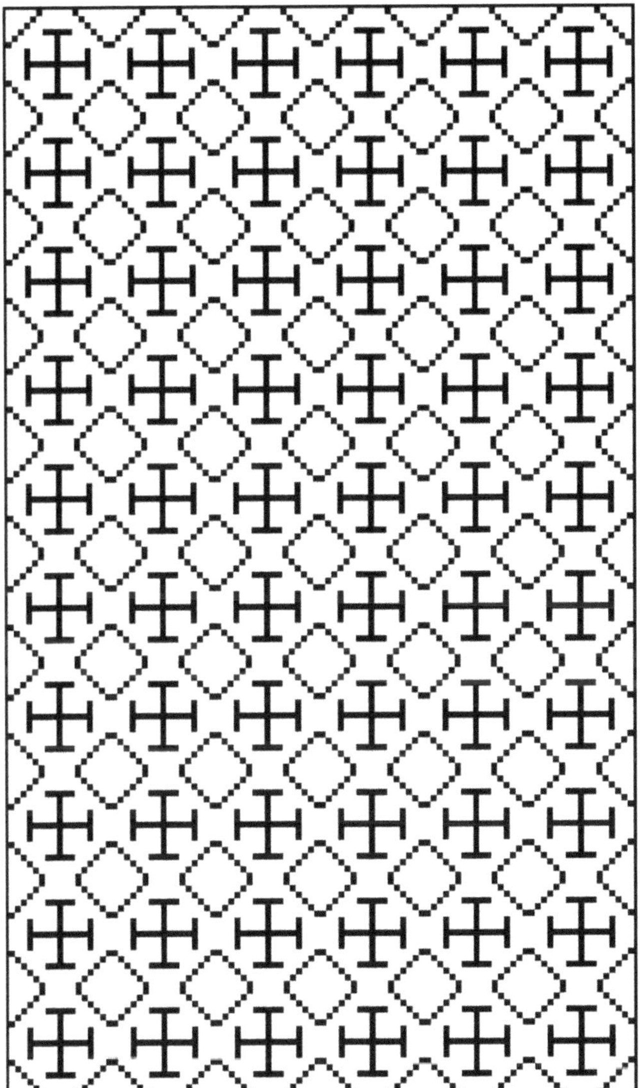

TSSS040SSS

Tile Patterns: Symmetry-Based Replication

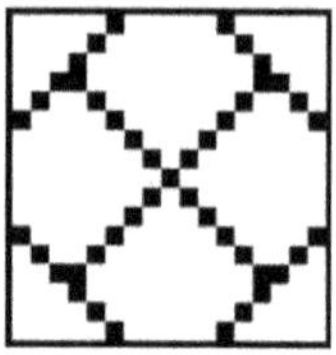

TSSS041

TSSS041SSS

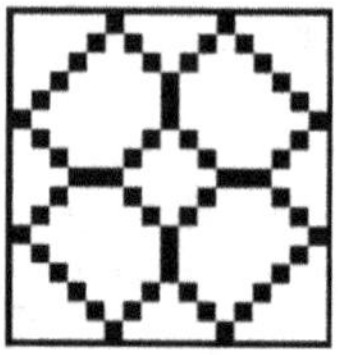

TSSS042

TSSS042SSS

Tile Patterns: Symmetry-Based Replication

TSSS043

TSSS043SSS

Tile Patterns: Symmetry-Based Replication

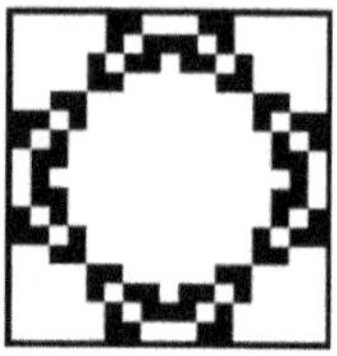

TSSS044

TSSS044SSS

Tile Patterns: Symmetry-Based Replication

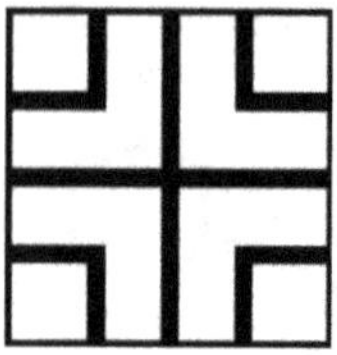

TSSS045

TSSS045SSS

Tile Patterns: Symmetry-Based Replication

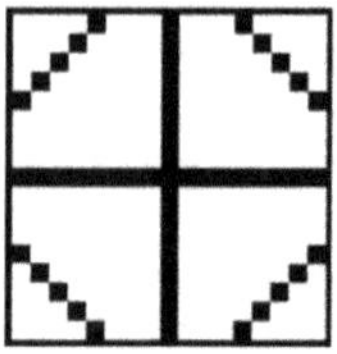

TSSS046

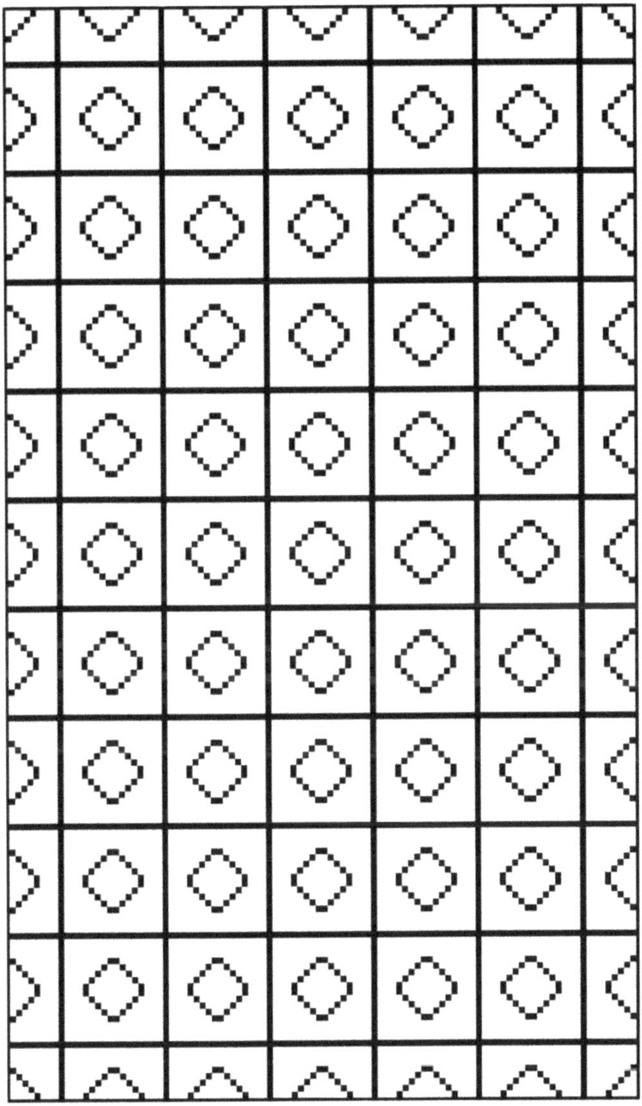

TSSS046SSS

Tile Patterns: Symmetry-Based Replication

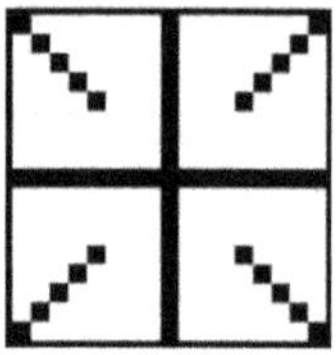

TSSS047

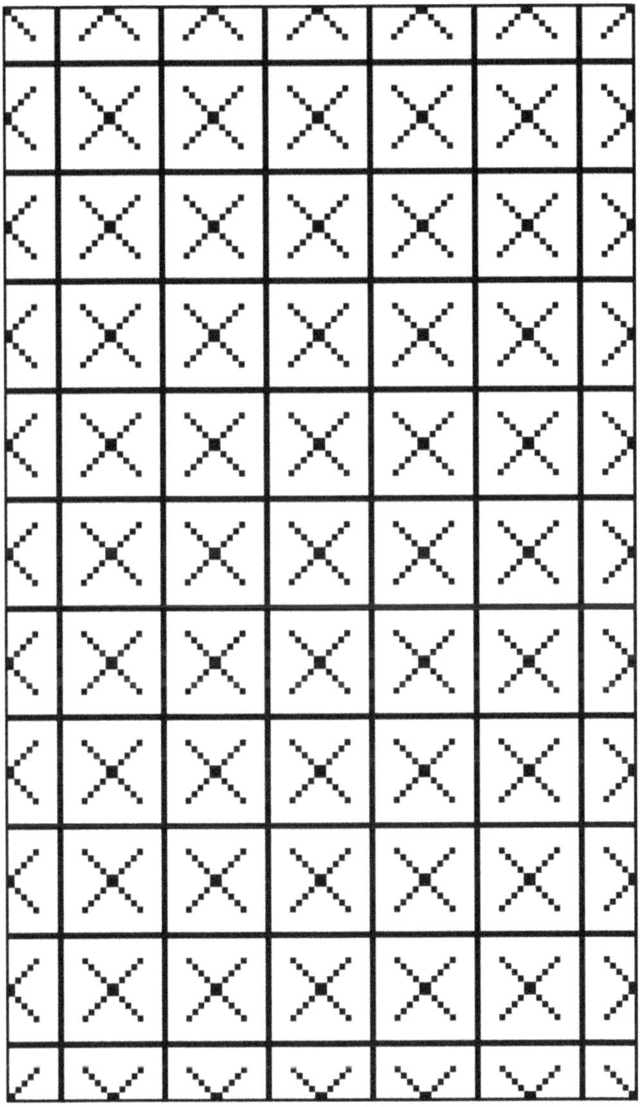

TSSS047SSS

Tile Patterns: Symmetry-Based Replication

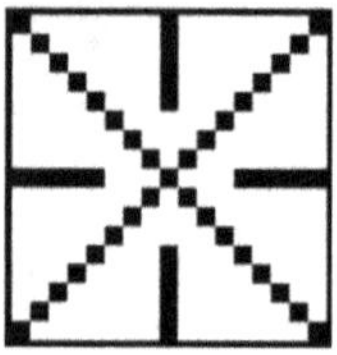

TSSS048

TSSS048SSS

Tile Patterns: Symmetry-Based Replication

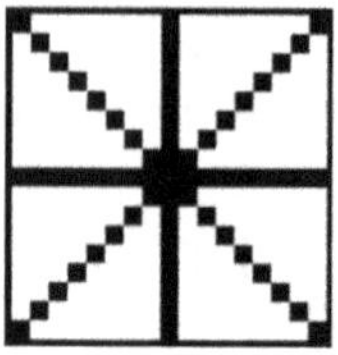

TSSS049

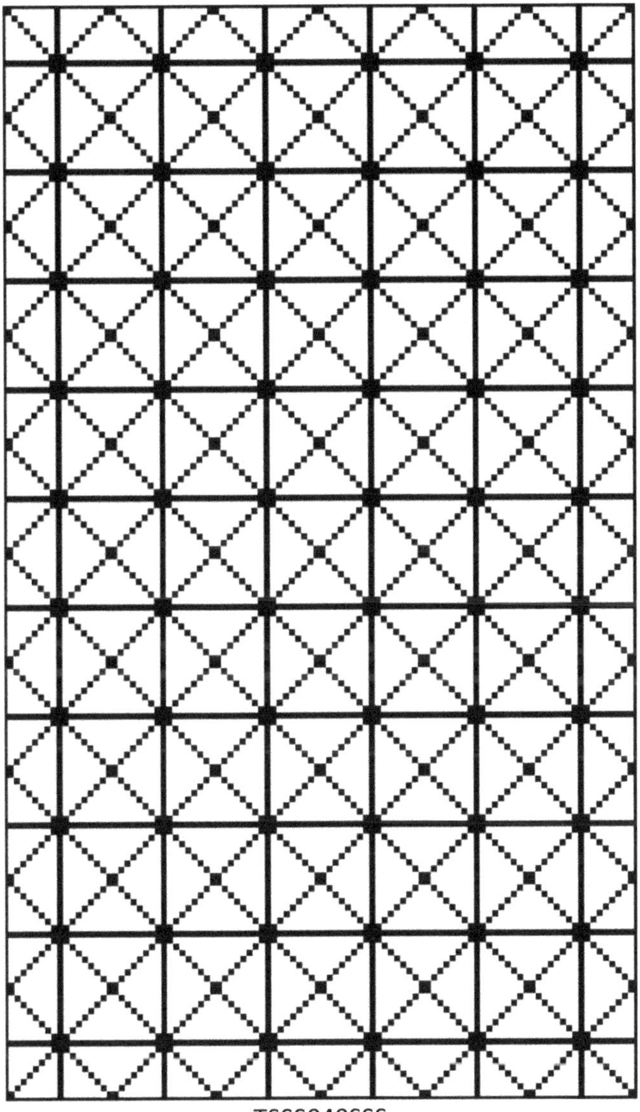

TSSS049SSS

Tile Patterns: Symmetry-Based Replication

TSSS050

TSSS050SSS

Tile Patterns: Symmetry-Based Replication

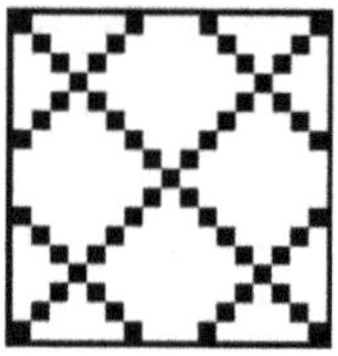

TSSS051

TSSS051SSS

TSSS052

TSSS052SSS

Tile Patterns: Symmetry-Based Replication

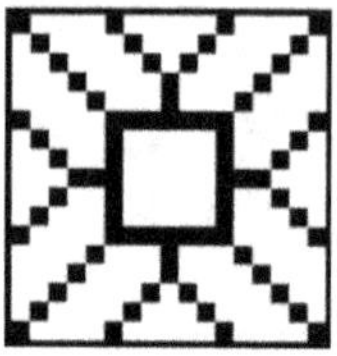

TSSS053

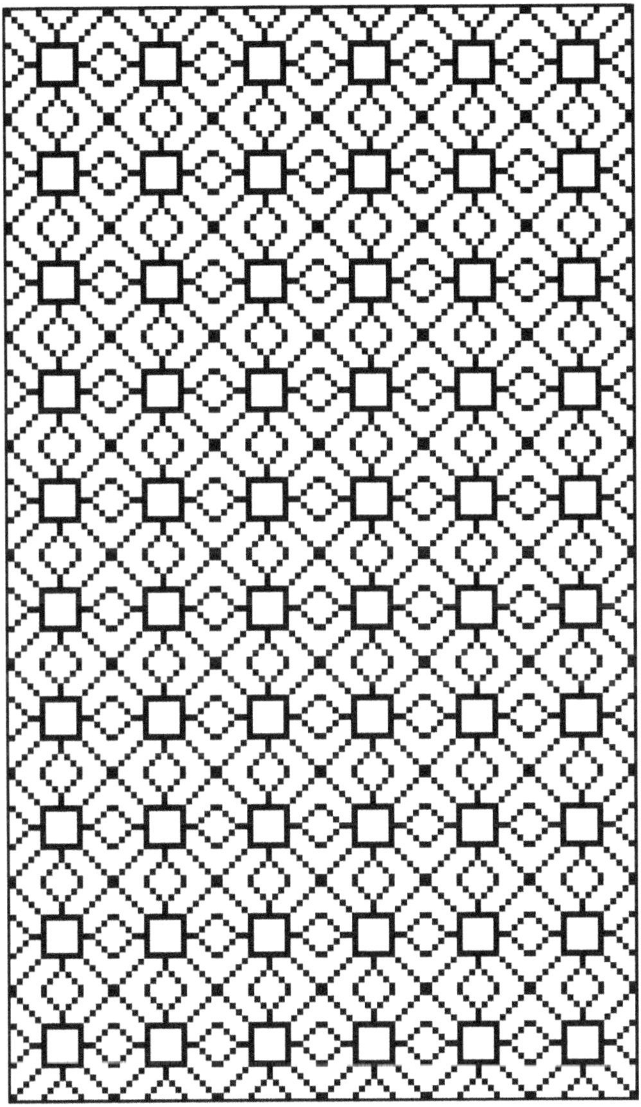

TSSS053SSS

Tile Patterns: Symmetry-Based Replication

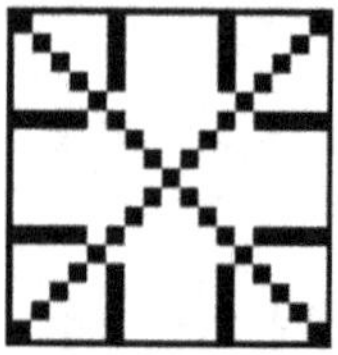

TSSS055

TSSS055SSS

Tile Patterns: Symmetry-Based Replication

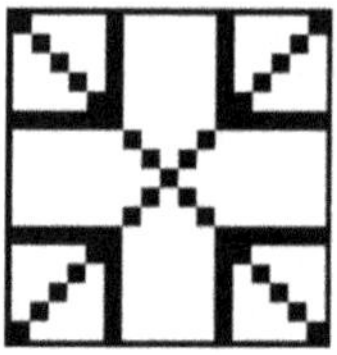

TSSS056

TSSS056SSS

Tile Patterns: Symmetry-Based Replication

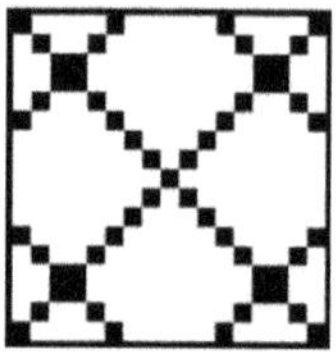

TSSS057

TSSS057SSS

Tile Patterns: Symmetry-Based Replication

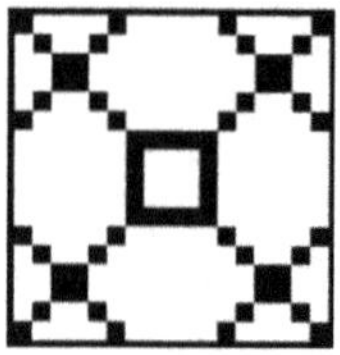

TSSS058

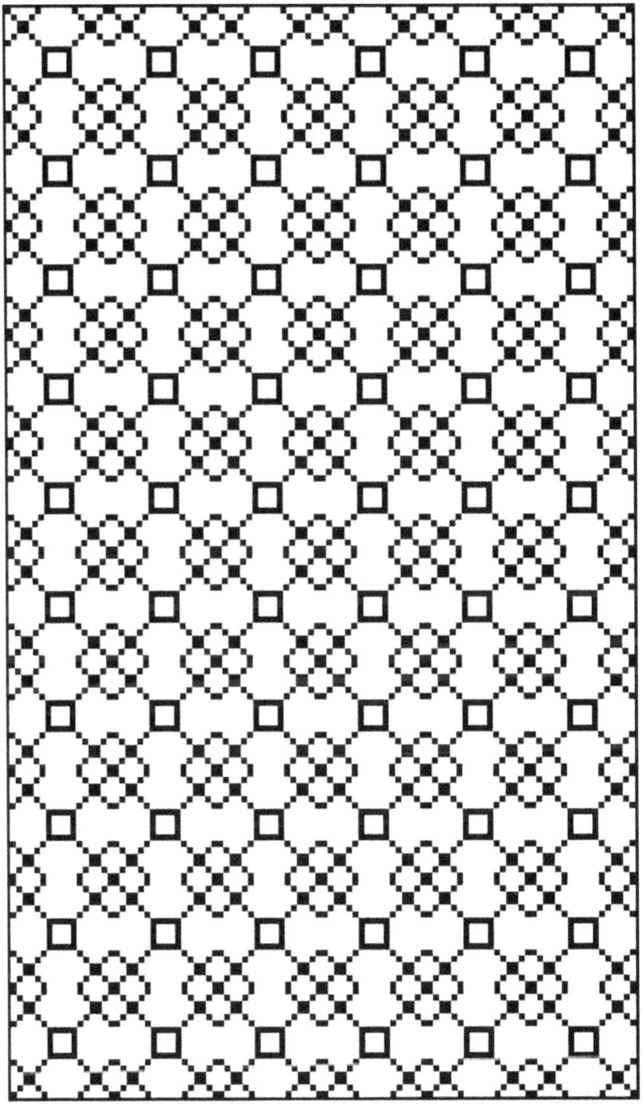

TSSS058SSS

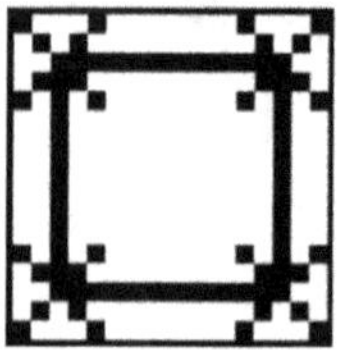

TSSS059

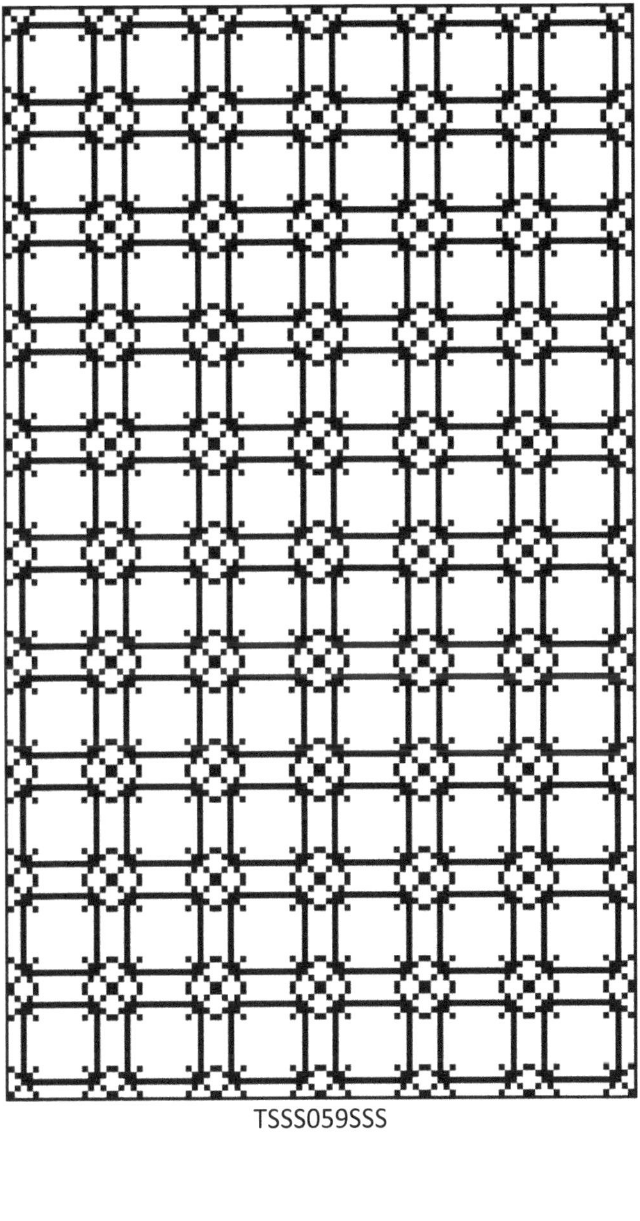

TSSS059SSS

Tile Patterns: Symmetry-Based Replication

TSSS060

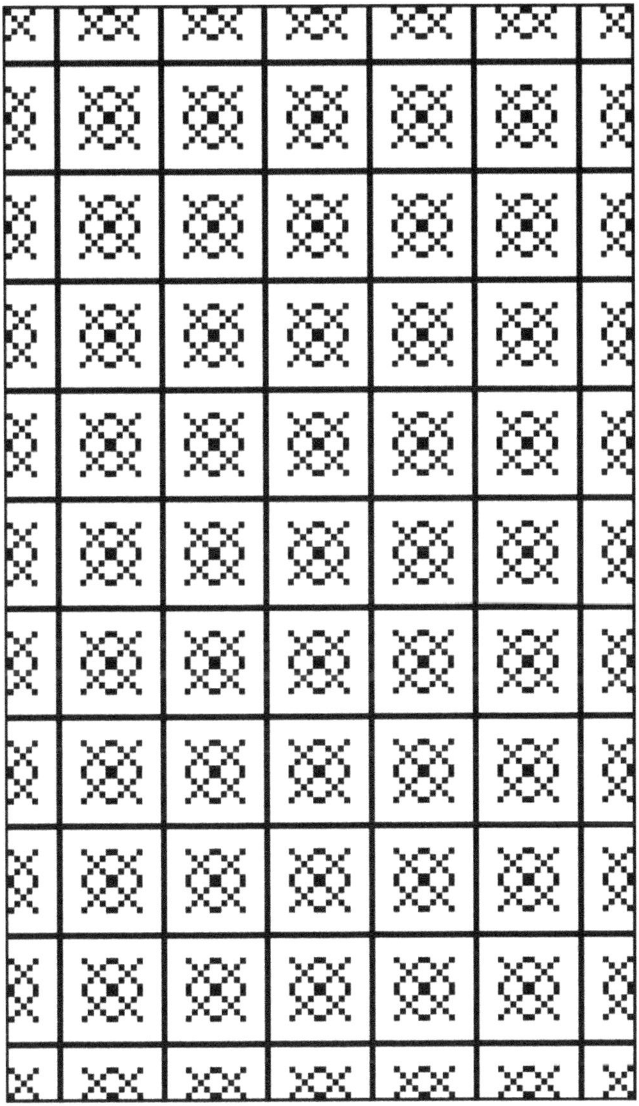

TSSS060SSS

Chapter 2: One Asymmetry

The simple patterns in this chapter exhibit one asymmetry: rotational or one reflection.

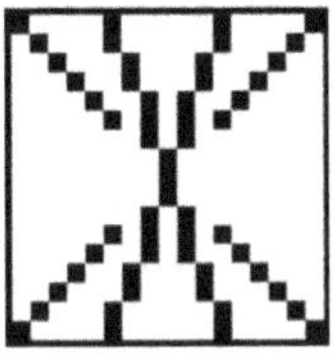

TSSA033

TSSA033SSS

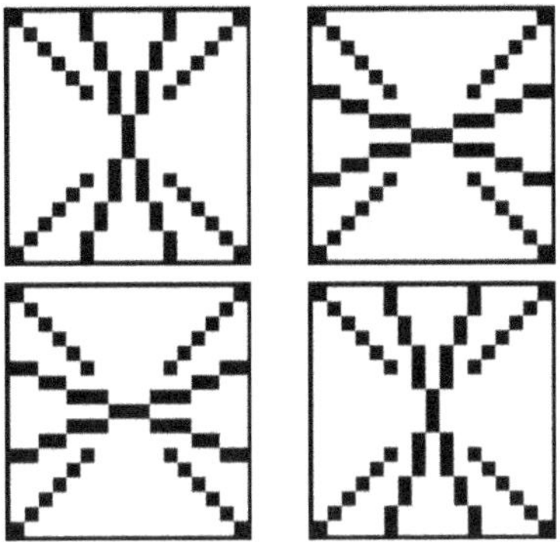

TSSA033 - Rotation

TSSA033SSA

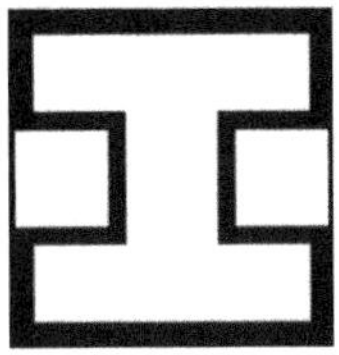

TSSA034

TSSA034SSS

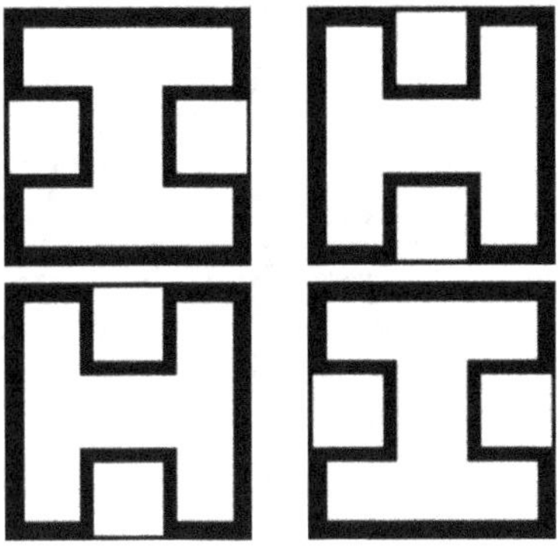

TSSA034 - Rotation

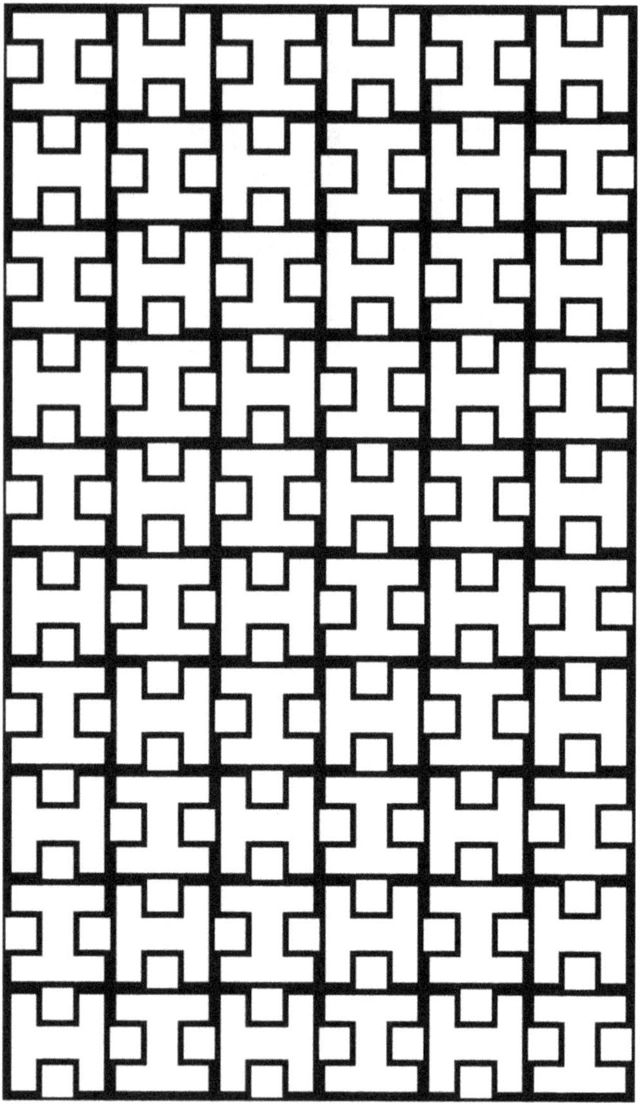

TSSA034SSA

Chapter 3: Two Asymmetries

The simple patterns in this chapter exhibit two asymmetries:

- Rotational and one of the reflections, or
- both reflections.

Tile Patterns: Symmetry-Based Replication

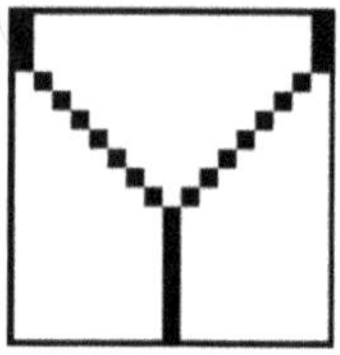

TSAA061

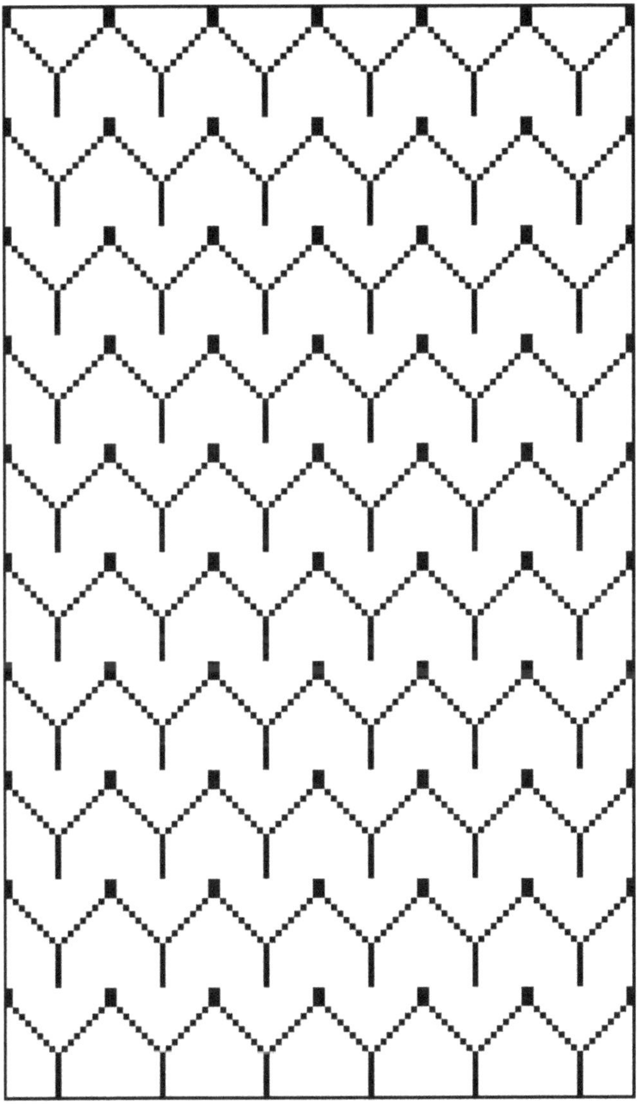

TSAA061SSS

Tile Patterns: Symmetry-Based Replication

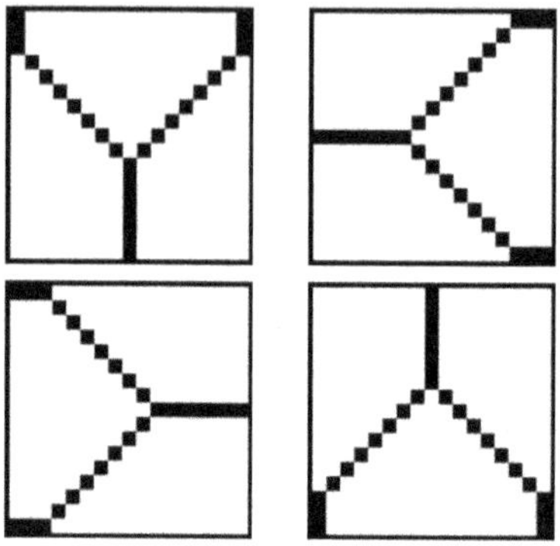

TSAA061 - Rotation

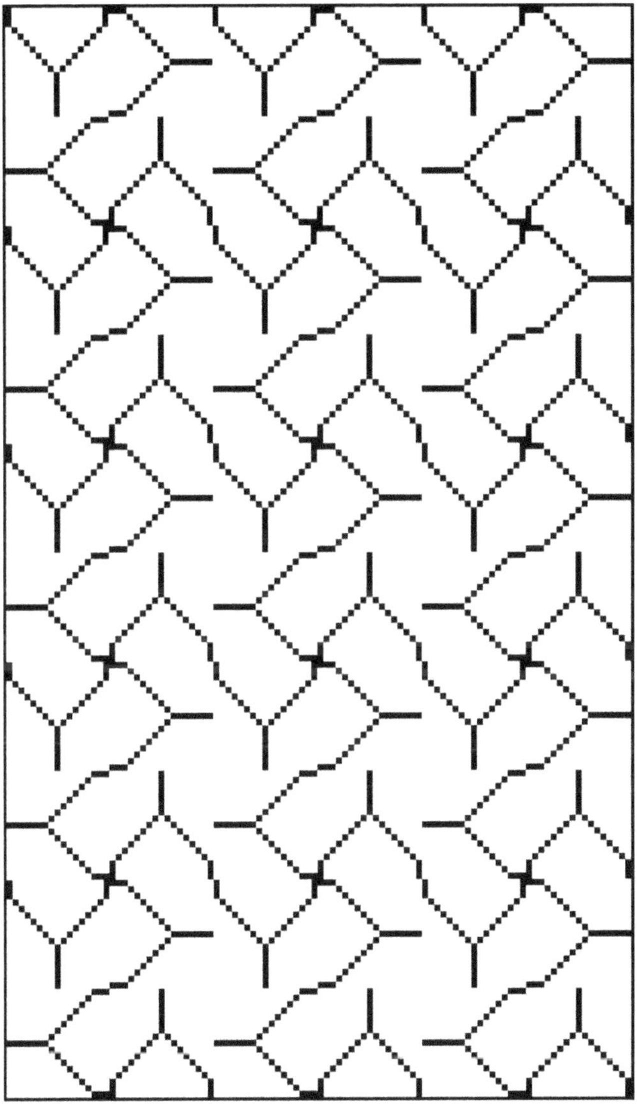

TSAA061SSA

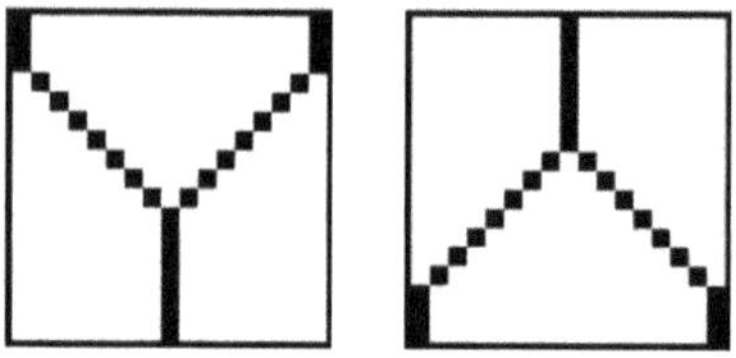

TSAA061 – Vertical Reflection

TSAA061SAS

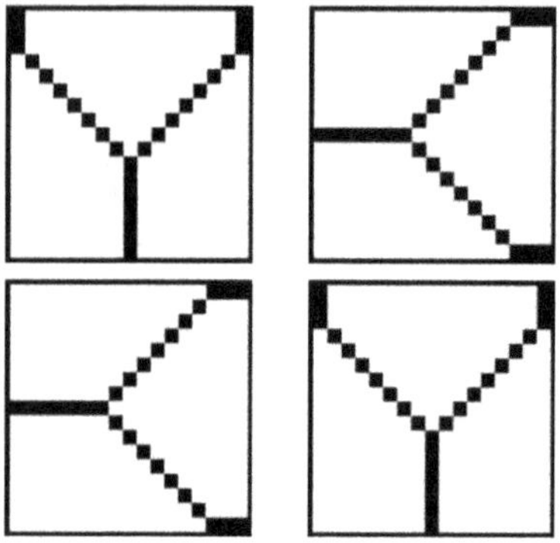

TSAA061 – Rotation and Vertical Reflection

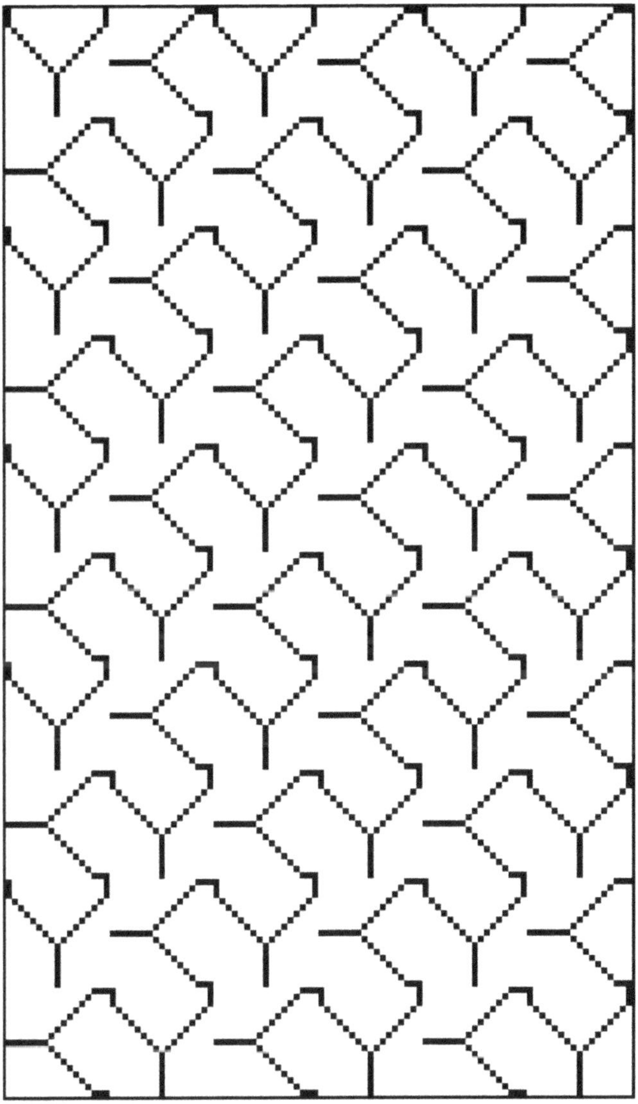

TSAA061SAA

Tile Patterns: Symmetry-Based Replication

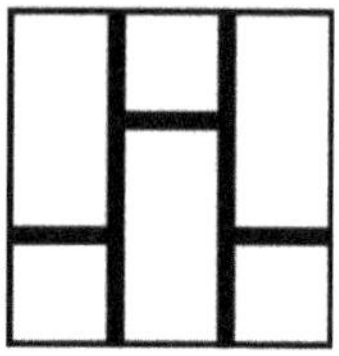

TASA032

TASA032SSS

Tile Patterns: Symmetry-Based Replication

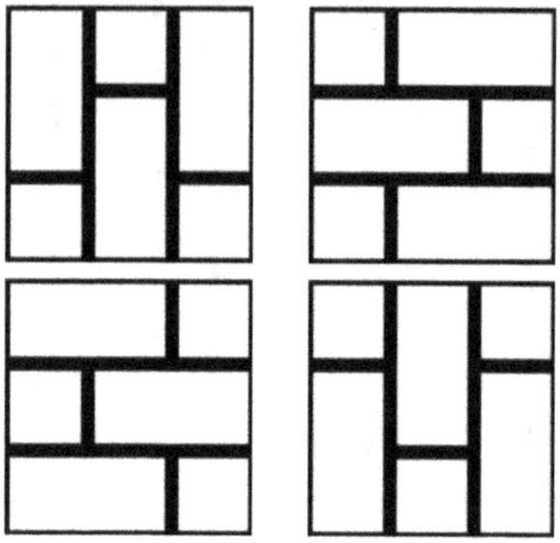

TASA032 - Rotation

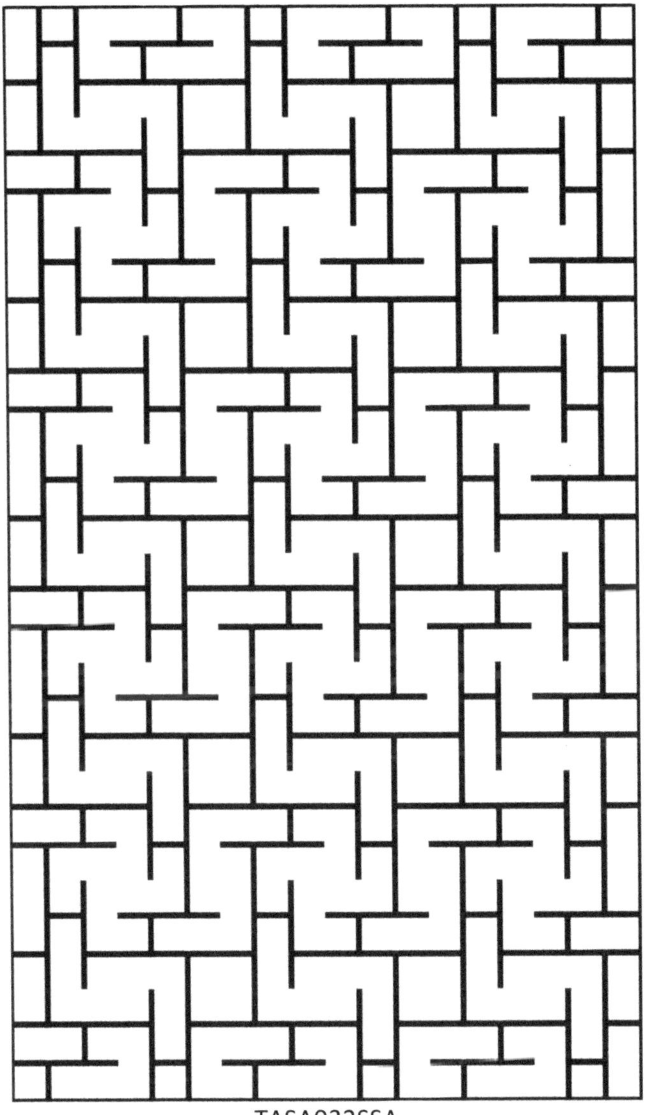

TASA032SSA

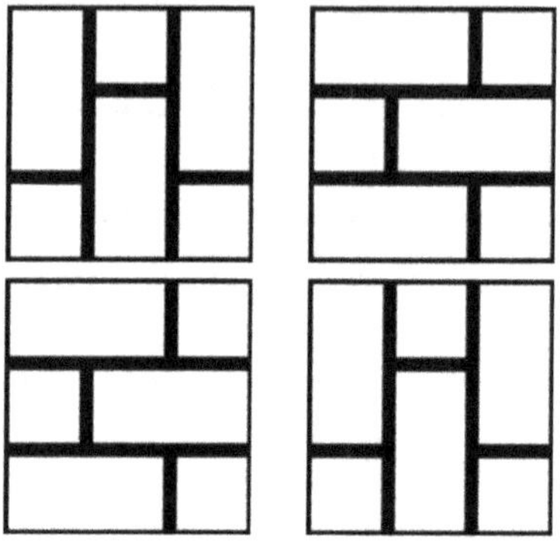

TASA032 – Rotation and Horizontal Reflection

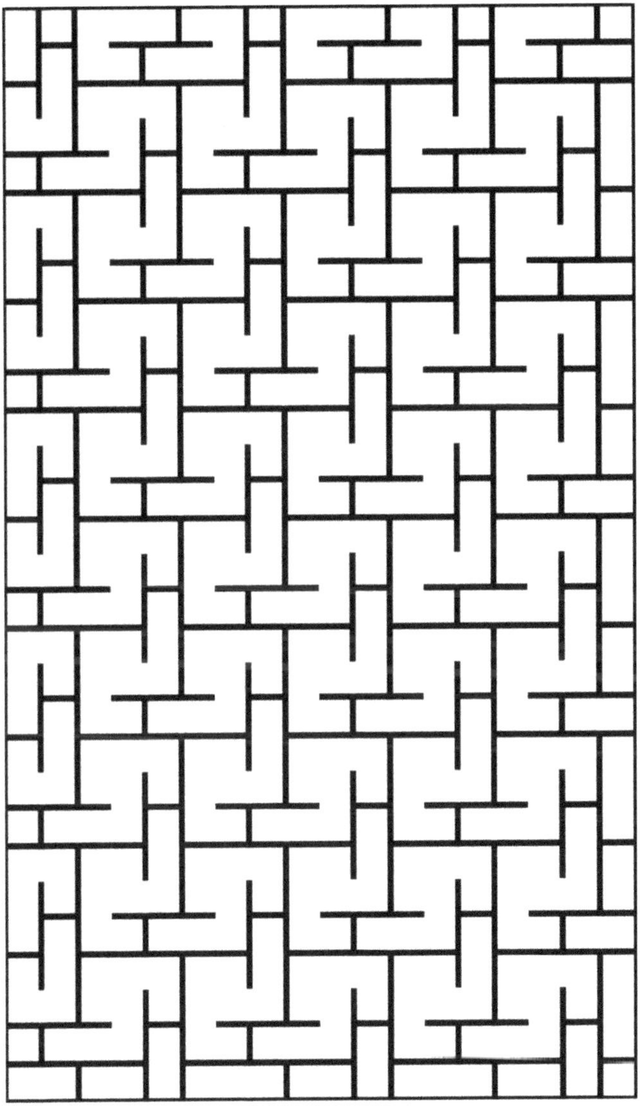

TASA032ASA

Tile Patterns: Symmetry-Based Replication

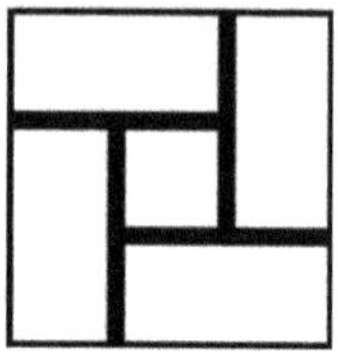

TAAS028

TAAS028SSS

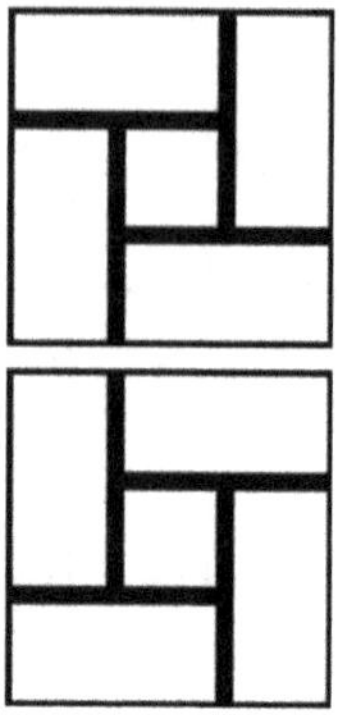

TAAS028 – Horizontal Reflection

TAAS028ASS

Tile Patterns: Symmetry-Based Replication

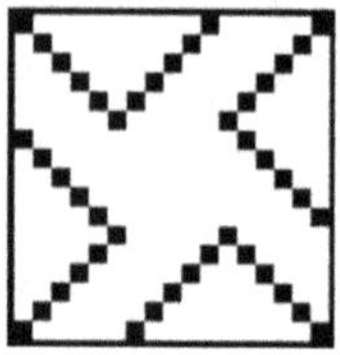

TAAS029

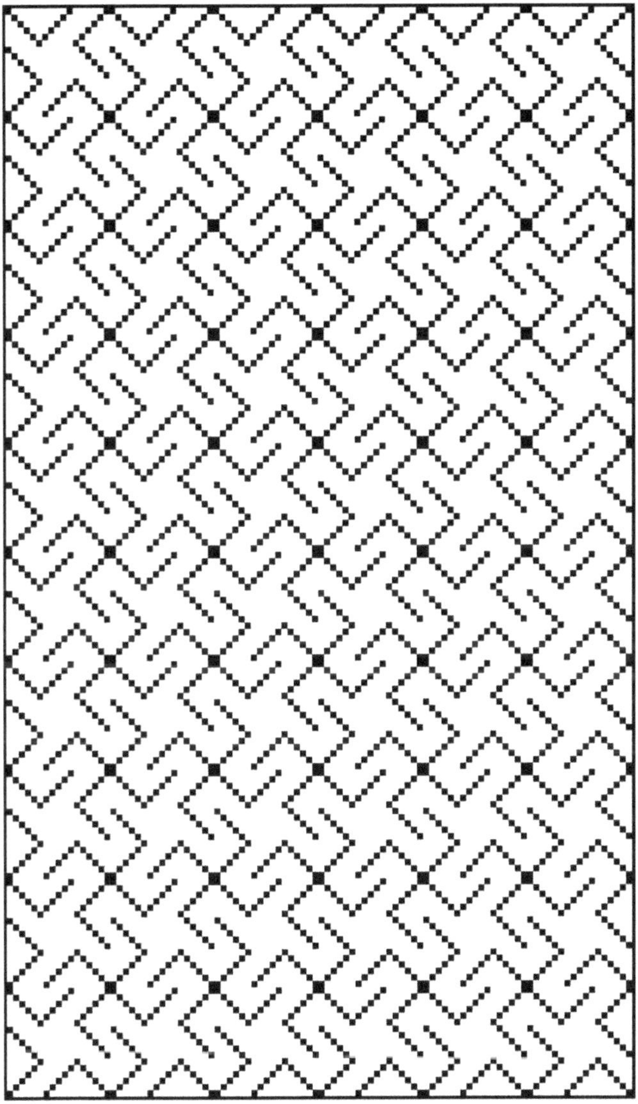

TAAS029SSS

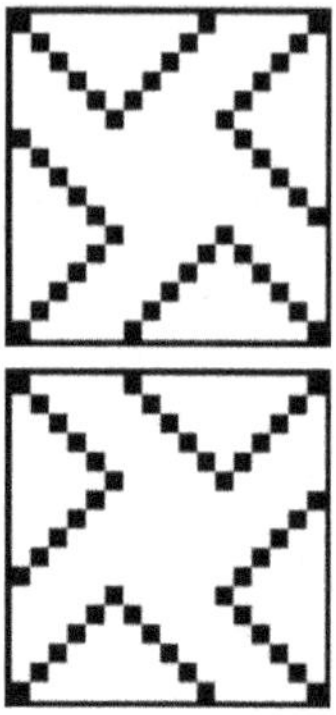

TAAS029 – Horizontal Reflection

TAAS029ASS

Tile Patterns: Symmetry-Based Replication

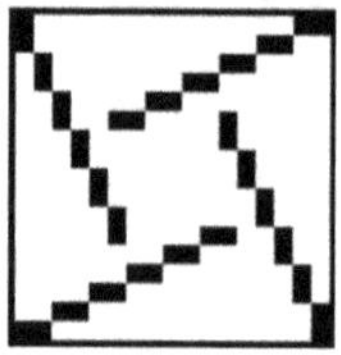

TAAS030

TAAS030SSS

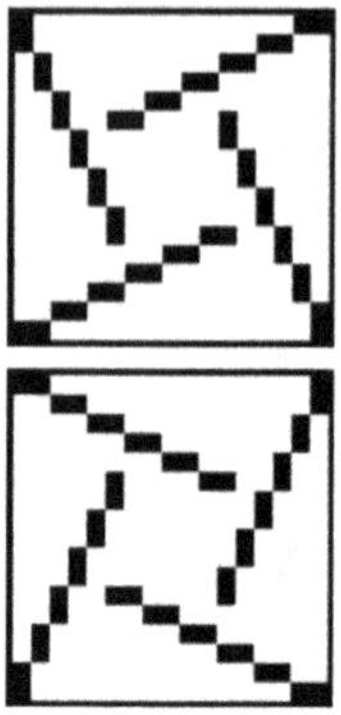

TAAS030 – Horizontal Reflection

TAAS030ASS

TAAS031

TAAS031SSS

TAAS031 – Horizontal Reflection

TAAS031ASS

Chapter 4: Three Asymmetries

The simple patterns in this chapter exhibit three asymmetries: rotational and both reflections. These simple patterns form the widest variety of tiled patterns because the combinations of transformations form different tile patterns.

Tile Patterns: Symmetry-Based Replication

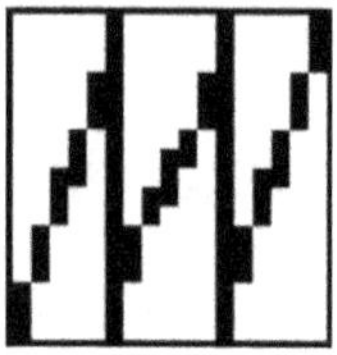

TAAA001

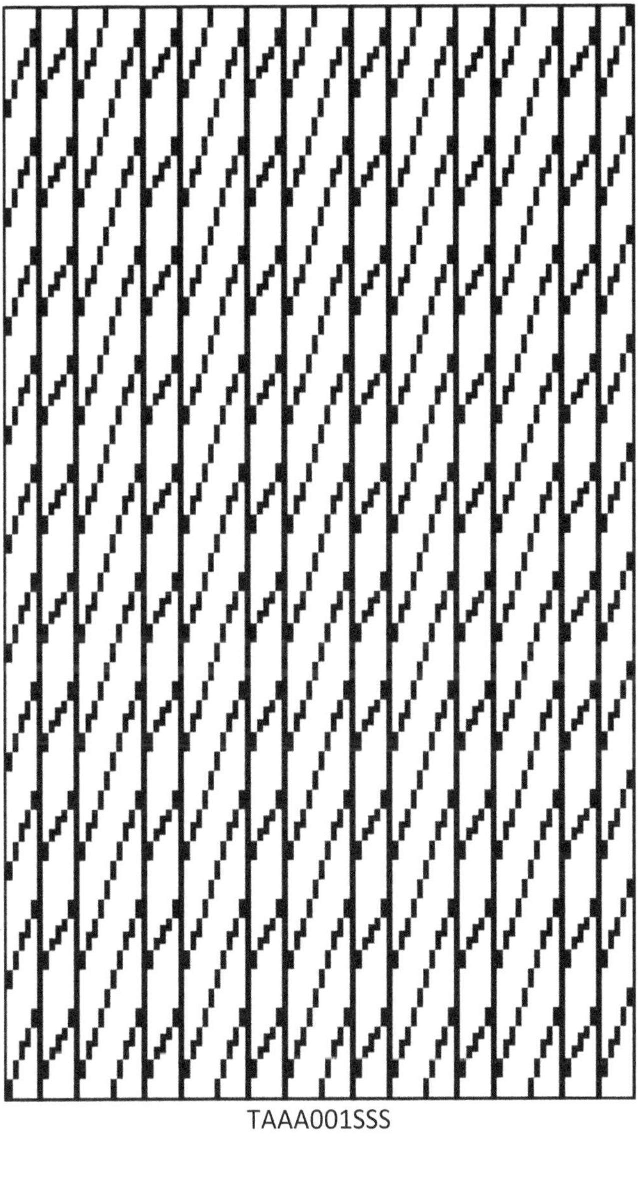

TAAA001SSS

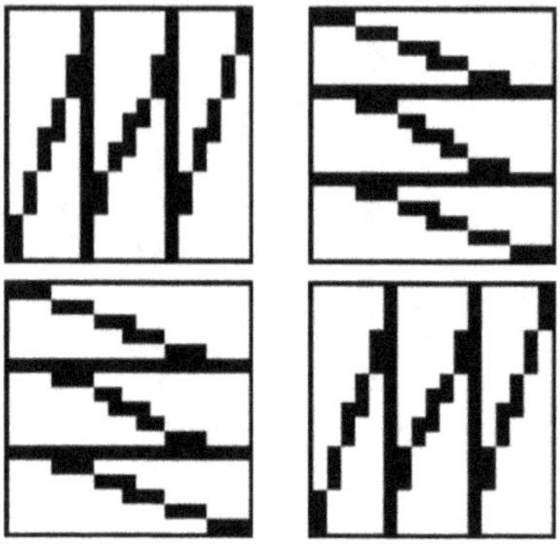

TAAA001 - Rotation

TAAA001SSA

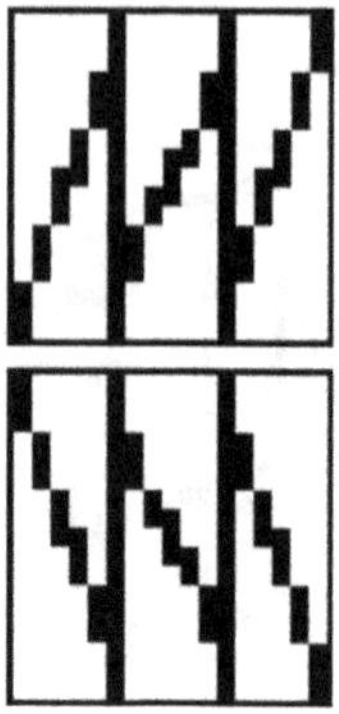

TAAA001 – Horizontal Reflection

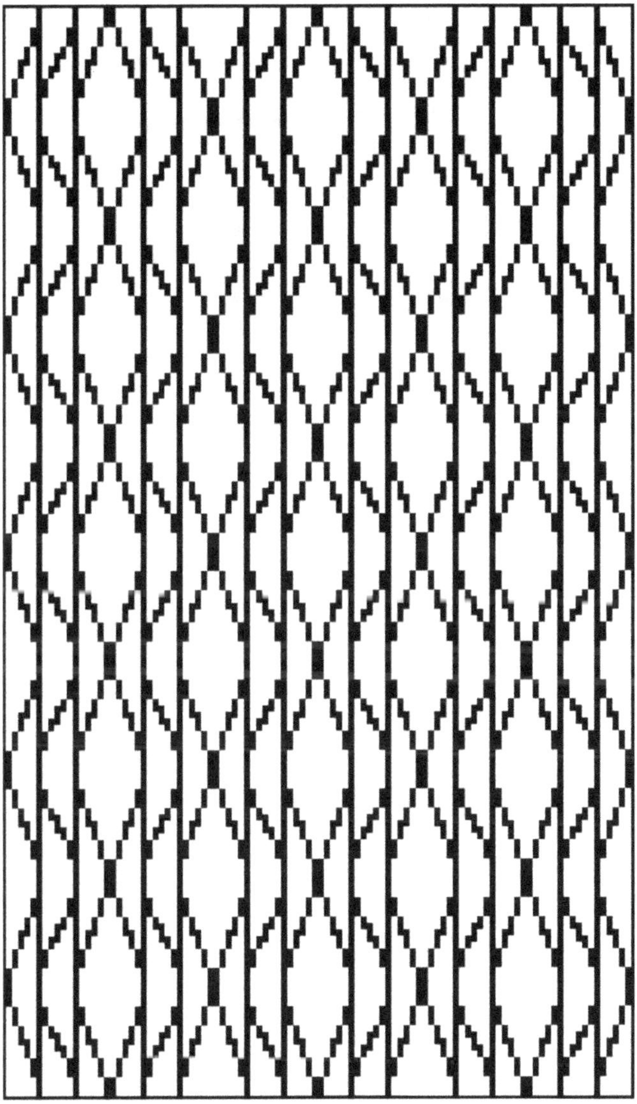

TAAA001ASS

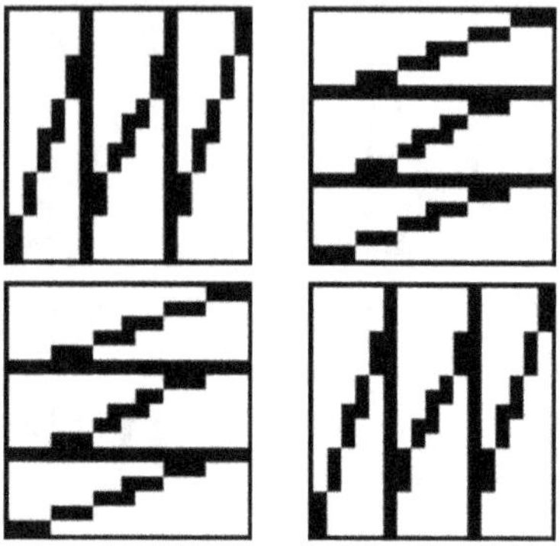

TAAA001 – Rotation and Vertical Reflection

TAAA001SAA

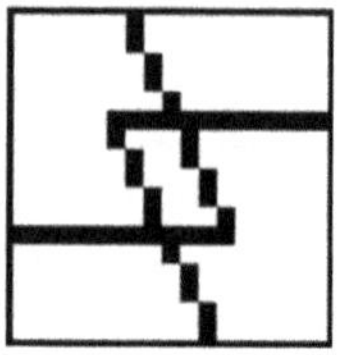

TAAA002

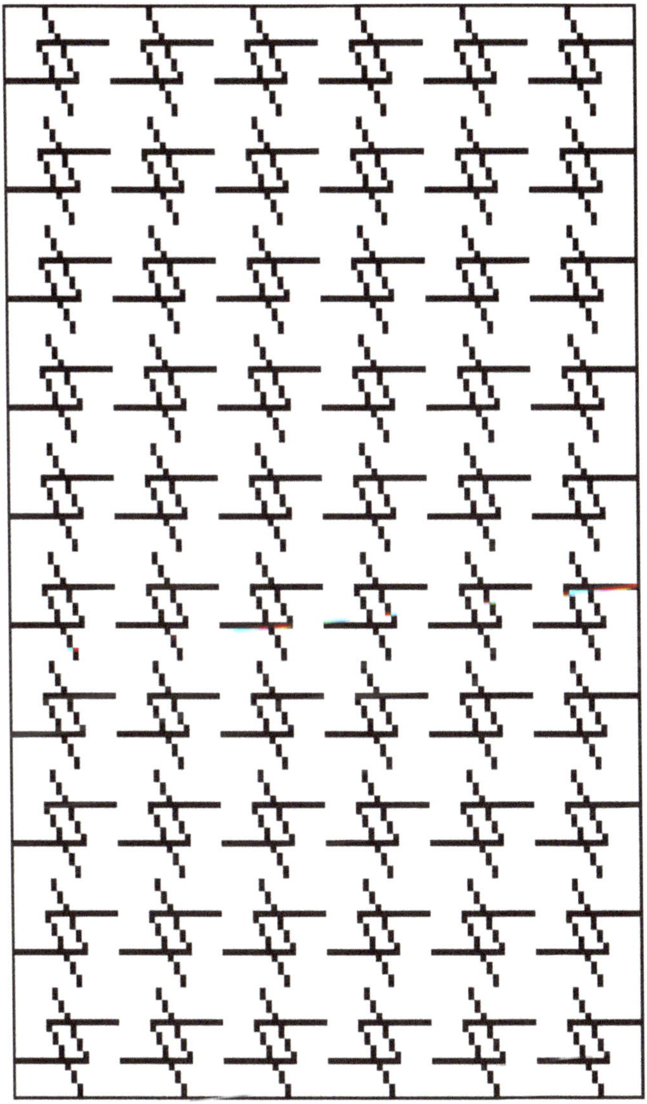

TAAA002SSS

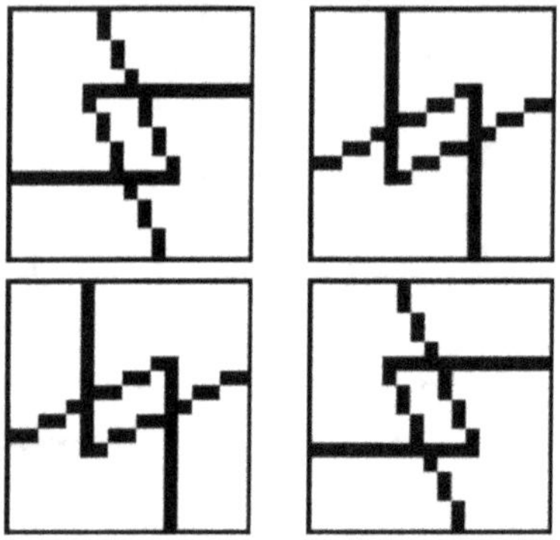

TAAA002 - Rotation

TAAA002SSA

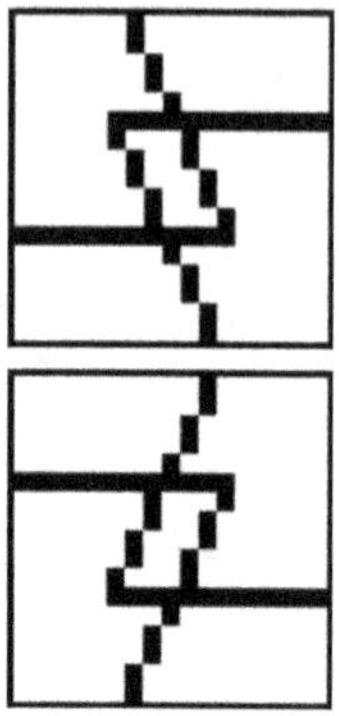

TAAA002 – Horizontal Reflection

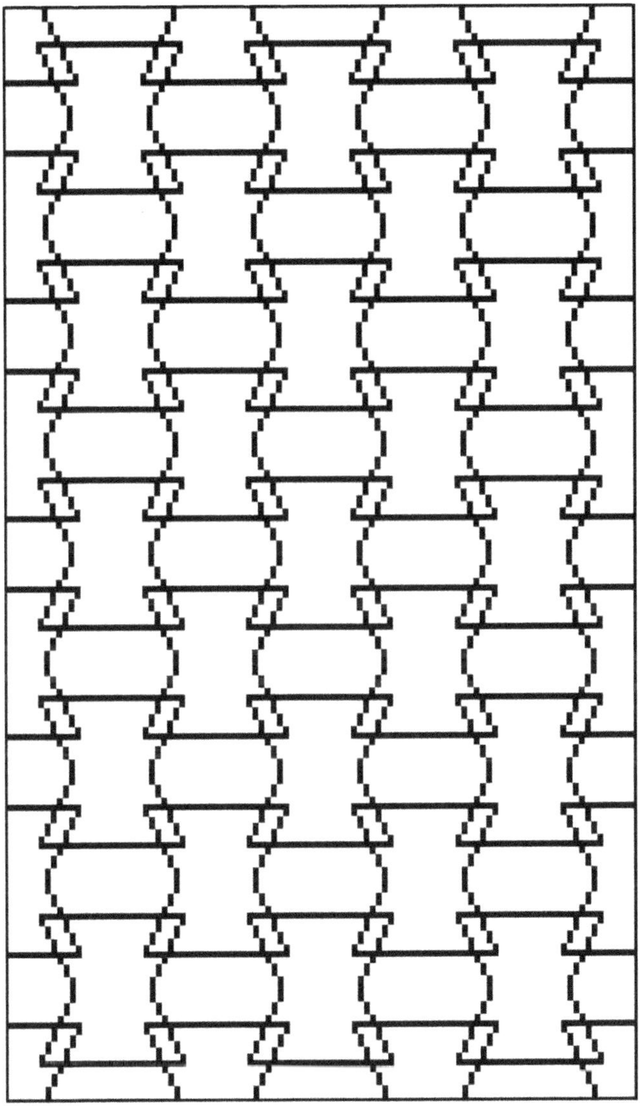

TAAA002ASS

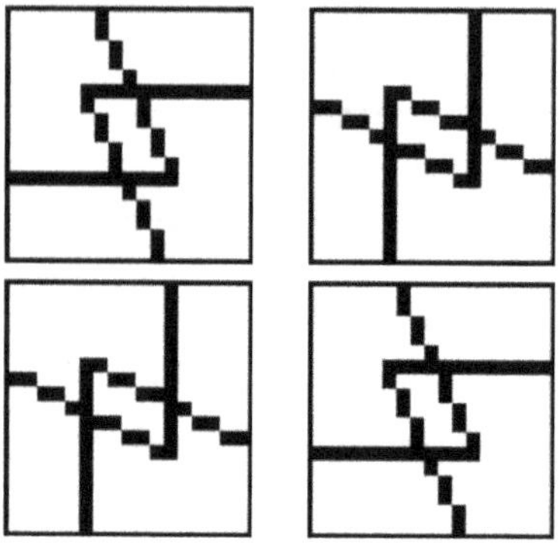

TAAA002 – Rotation and
Vertical Reflection

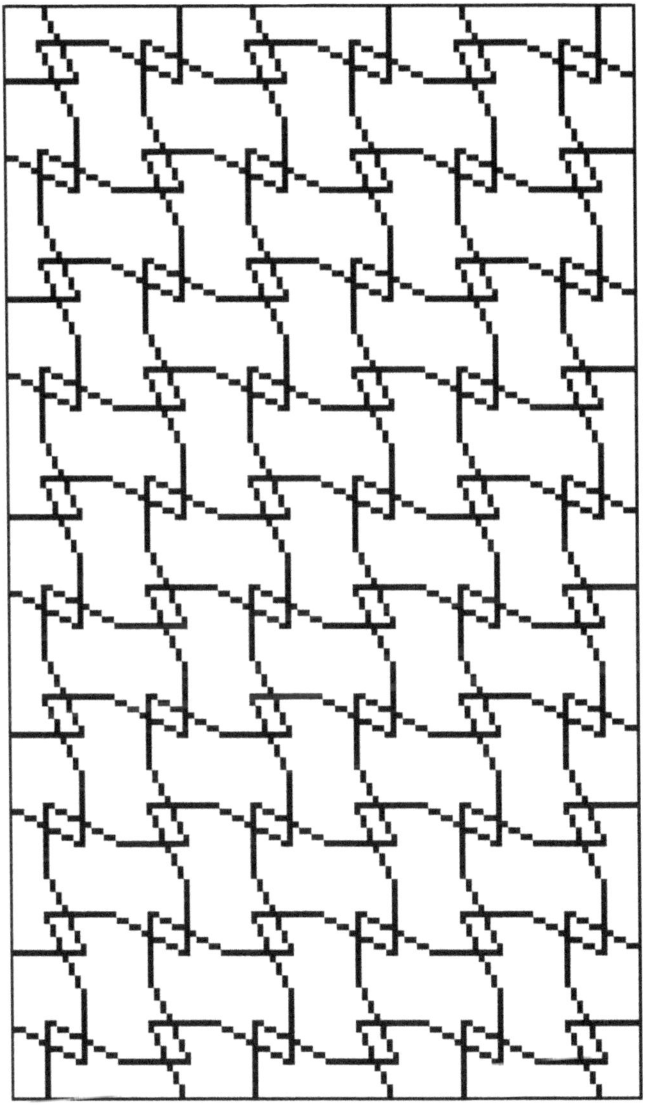

TAAA002SAA

Tile Patterns: Symmetry-Based Replication

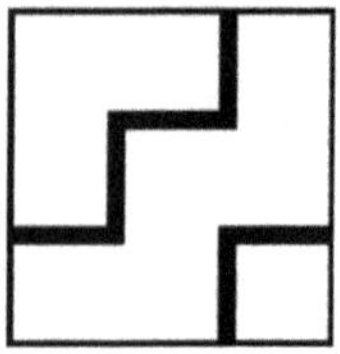

TAAA003

TAAA003SSS

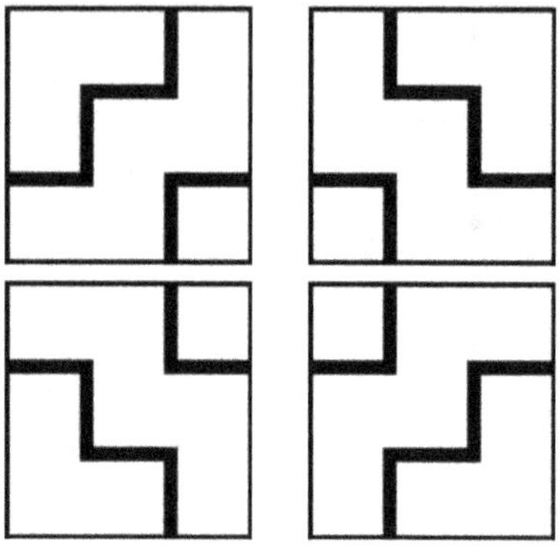

TAAA003 - Rotation

TAAA003SSA

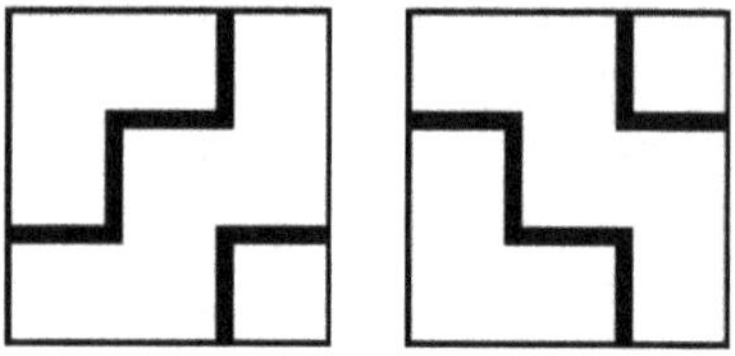

TAAA003 – Vertical Reflection

TAAA003SAS

Tile Patterns: Symmetry-Based Replication

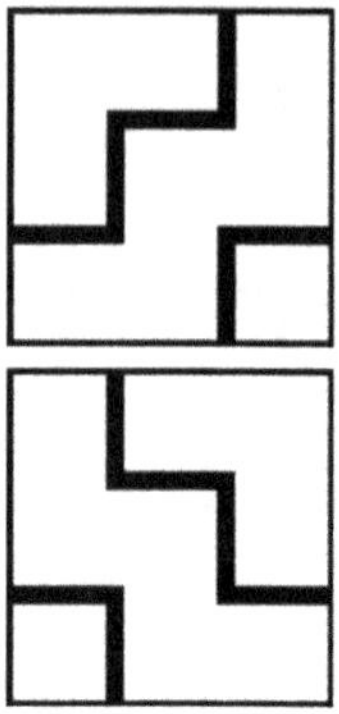

TAAA003 – Horizontal Reflection

TAAA003ASS

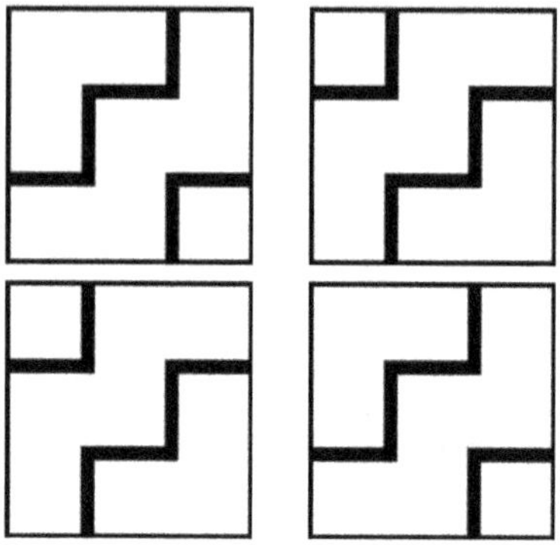

TAAA003 – Rotation and Vertical Reflection

TAAA003SAA

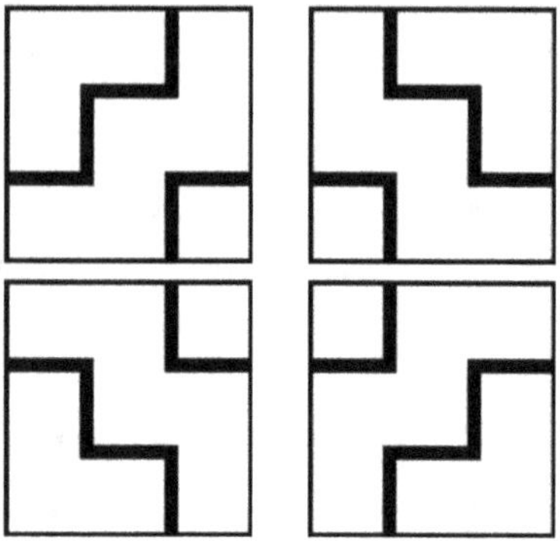

TAAA003 – Vertical and Horizontal Reflections

TAAA003AAS

Tile Patterns: Symmetry-Based Replication

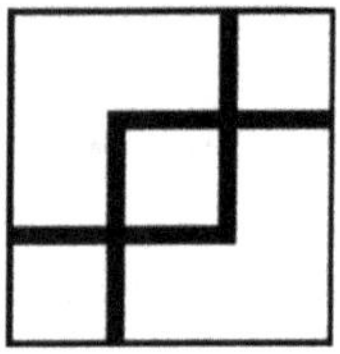

TAAA004

TAAA004SSS

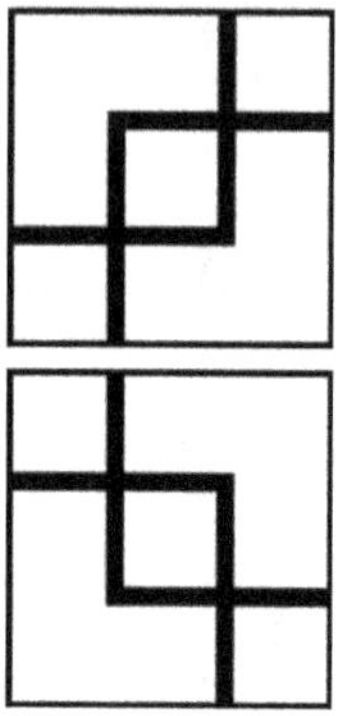

TAAA004 – Horizontal Reflection

TAAA004ASS

Tile Patterns: Symmetry-Based Replication

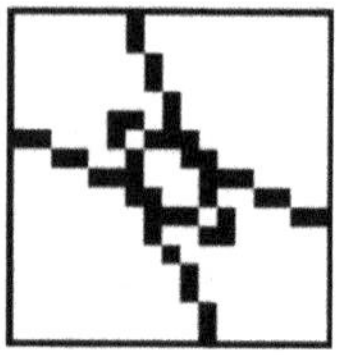

TAAA005

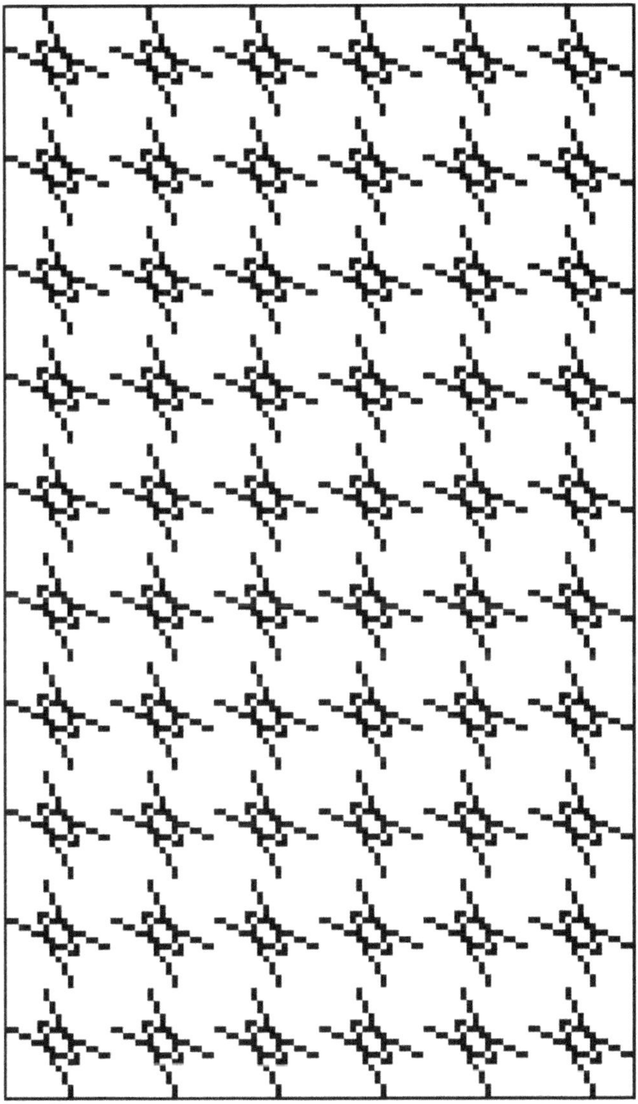

TAAA005SSS

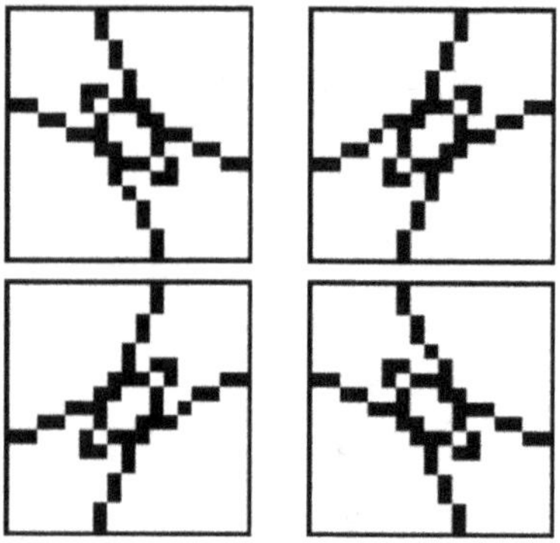

TAAA005 - Rotation

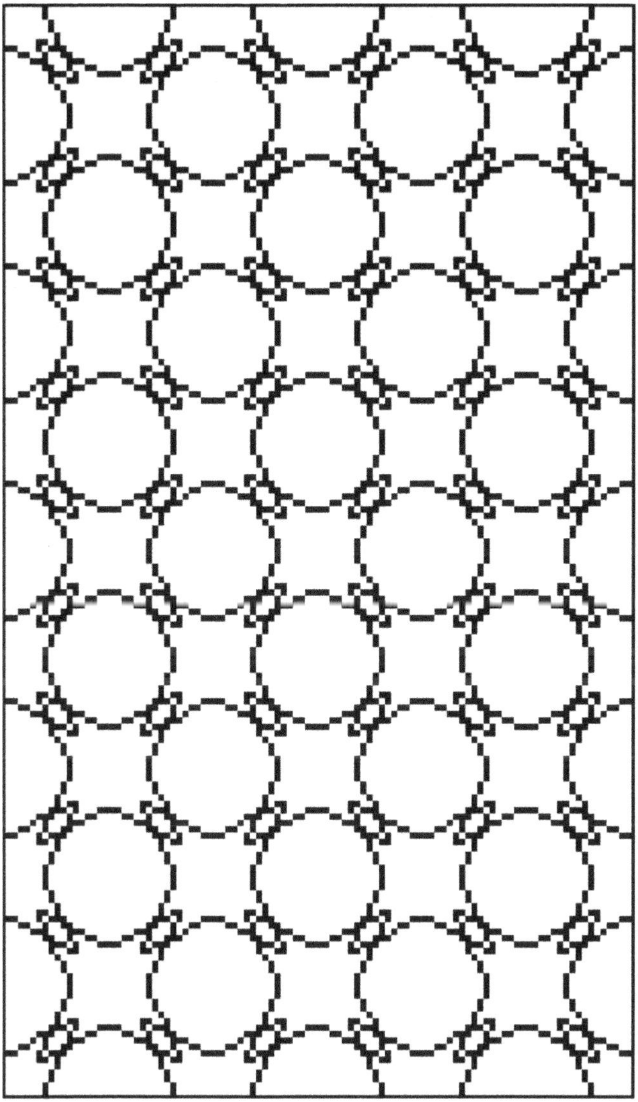

TAAA005SSA

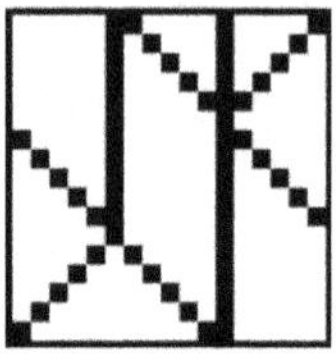

TAAA006

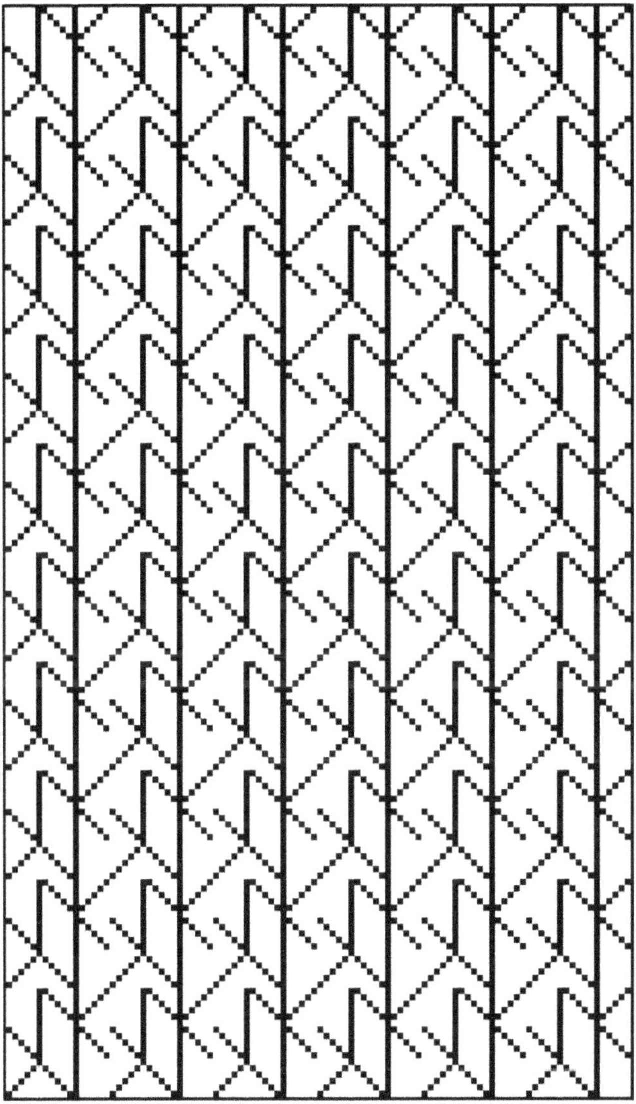

TAAA006SSS

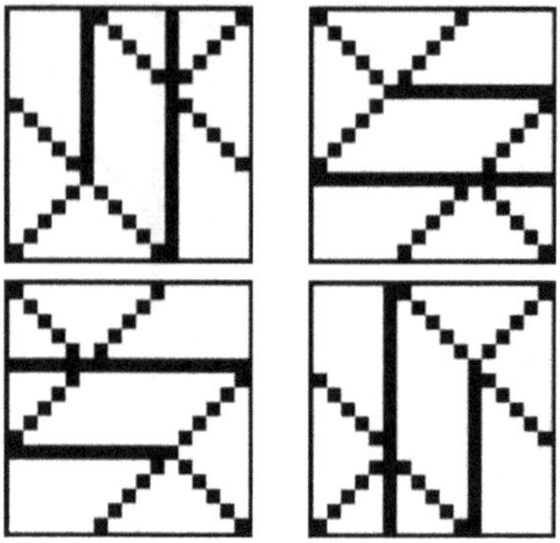

TAAA006 - Rotation

TAAA006SSA

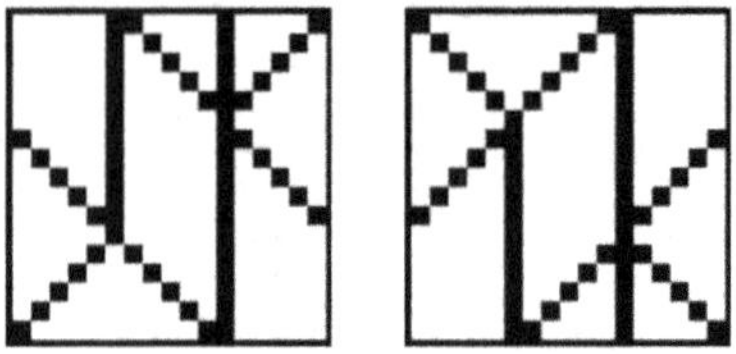

TAAA006 – Vertical Reflection

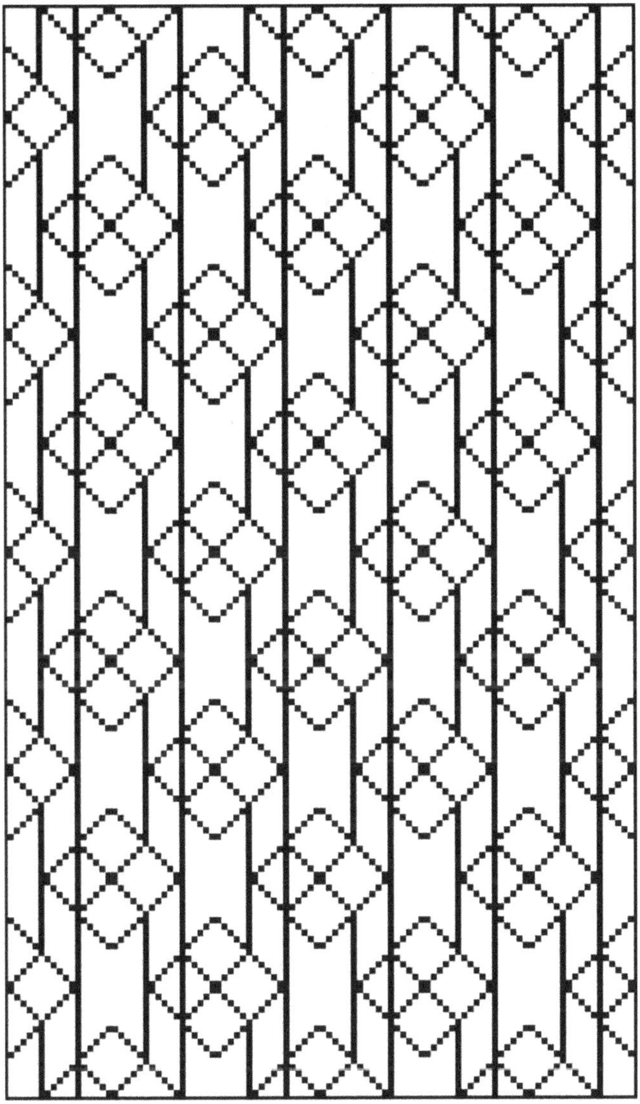

TAAA006SAS

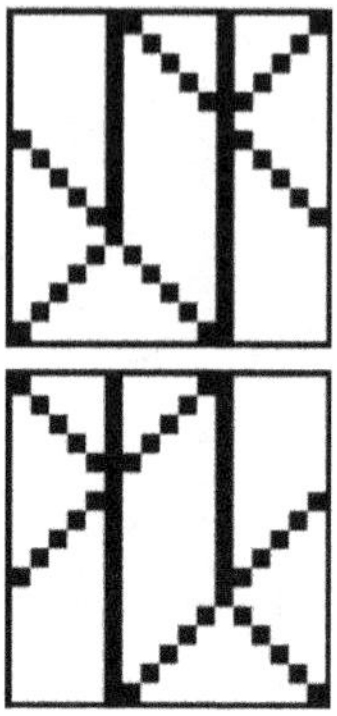

TAAA006 – Horizontal Reflection

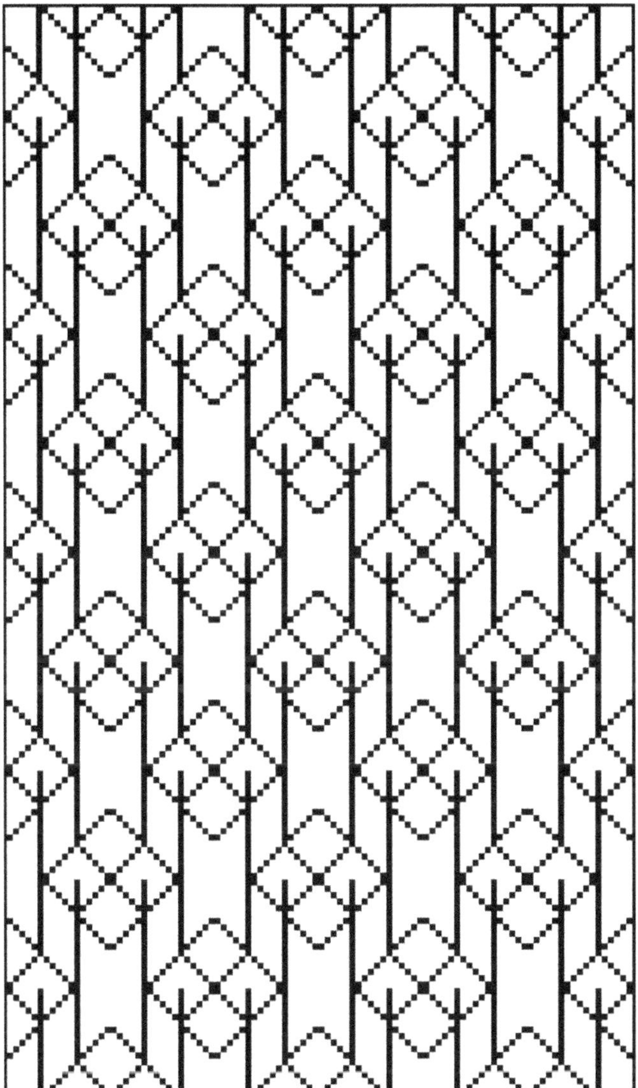

TAAA006ASS

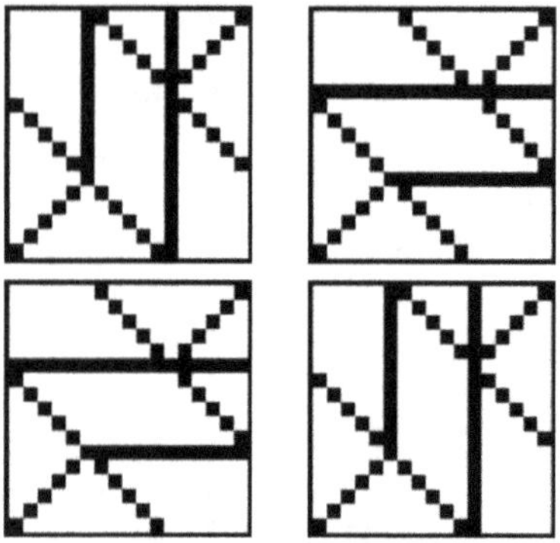

TAAA006 – Rotation and Vertical Reflection

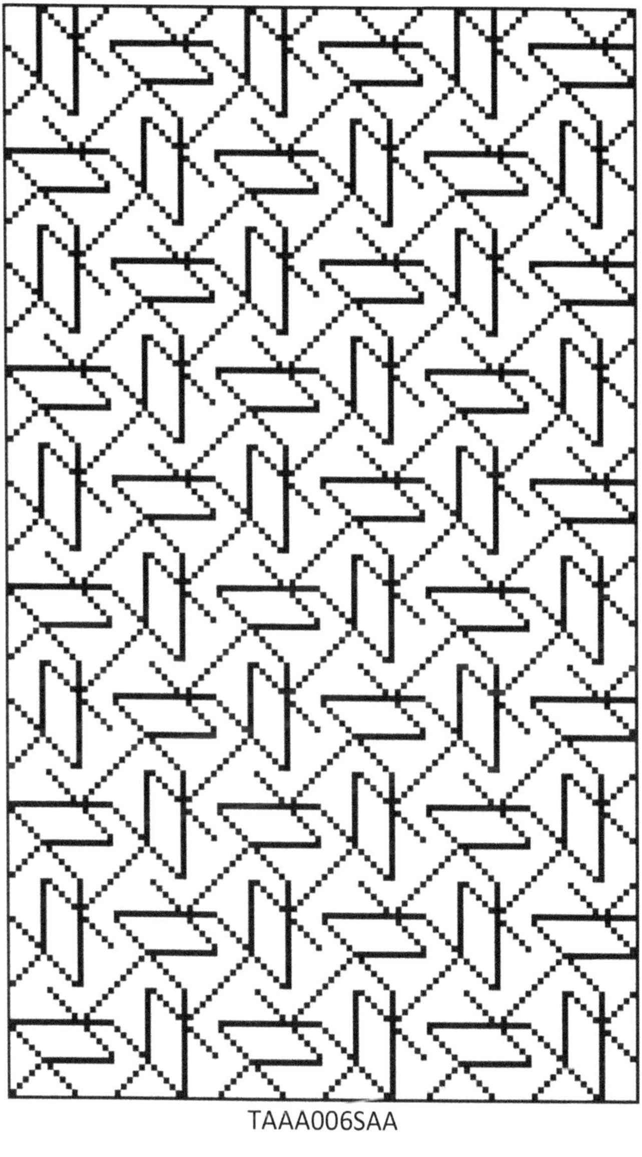

TAAA006SAA

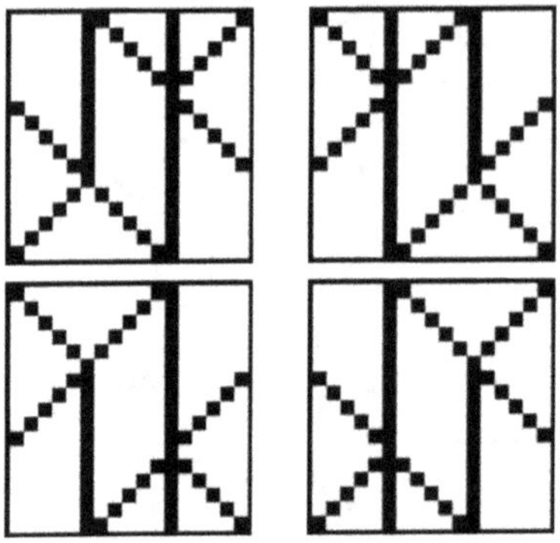

TAAA006 – Vertical and Horizontal Reflections

TAAA006AAS

TAAA007

TAAA007SSS

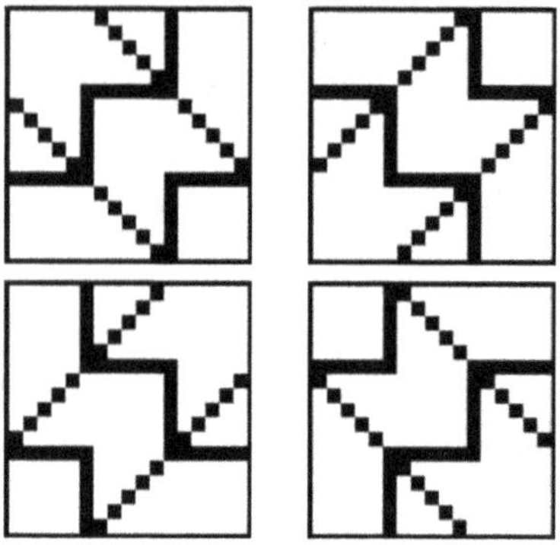

TAAA007 - Rotation

TAAA007SSA

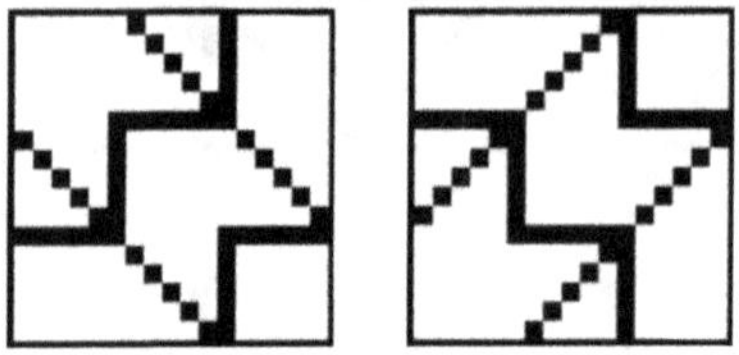

TAAA007 – Vertical Reflection

TAAA007SAS

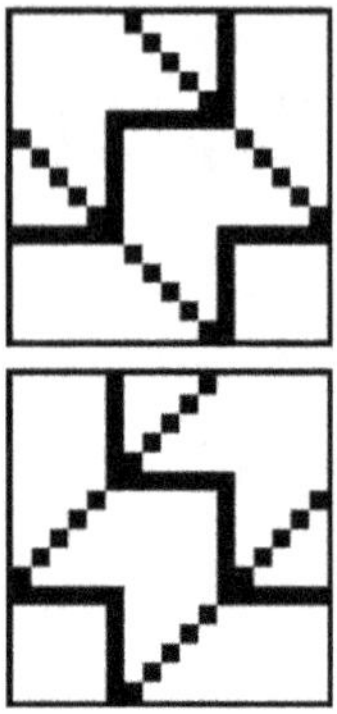

TAAA007 – Horizontal Reflection

TAAA007ASS

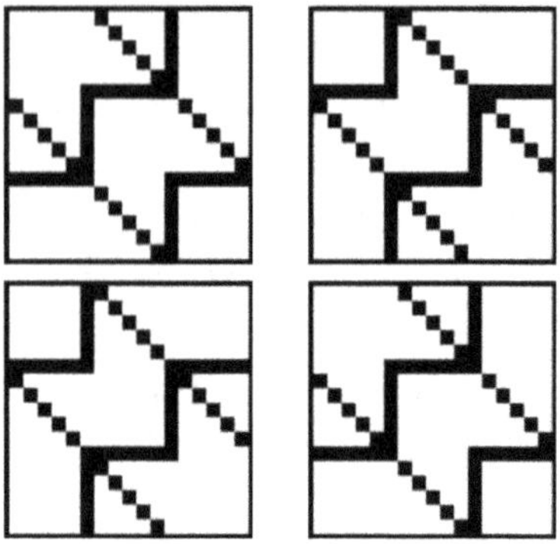

TAAA007 – Rotation and
Vertical Reflection

TAAA007SAA

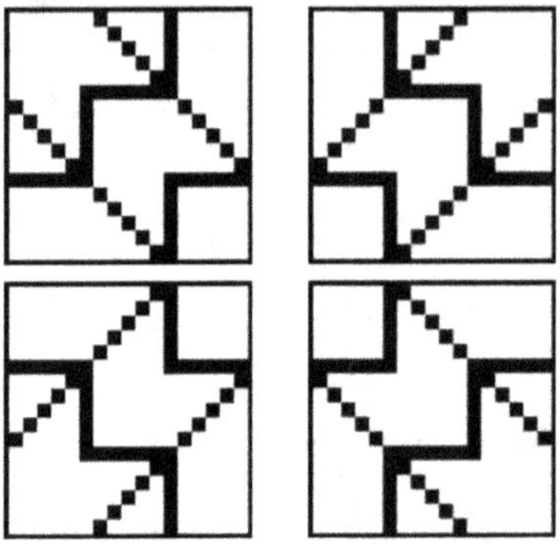

TAAA007 – Vertical and Horizontal Reflections

TAAA007AAS

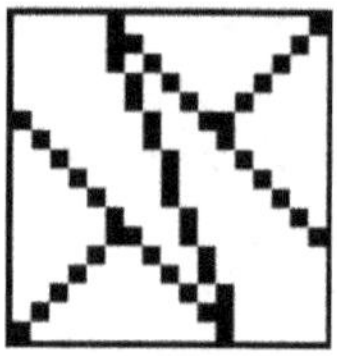

TAAA008

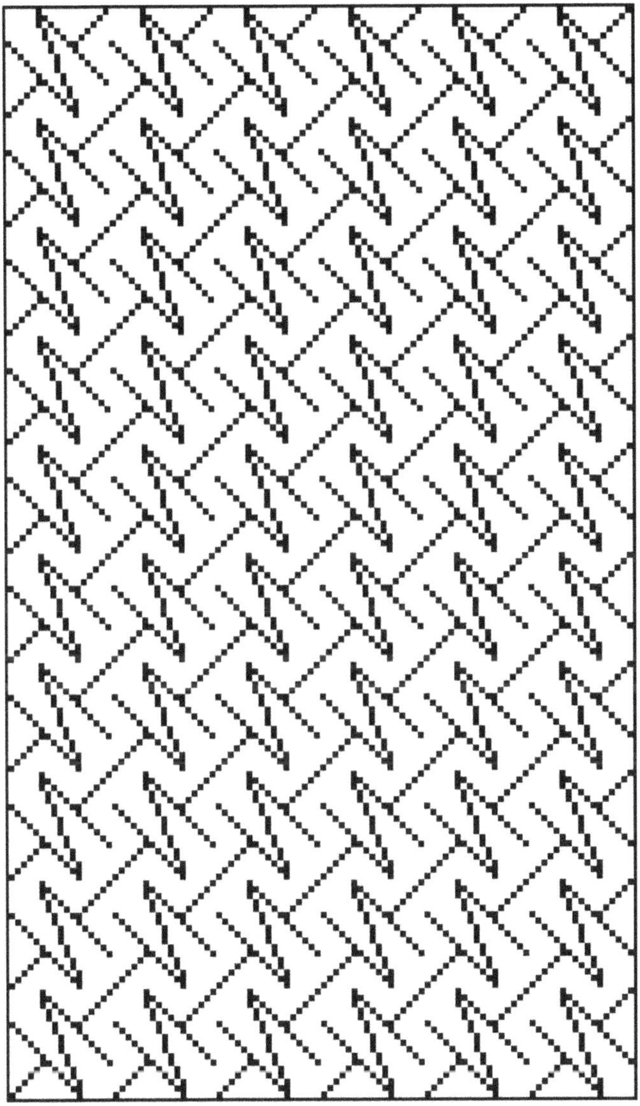

TAAA008SSS

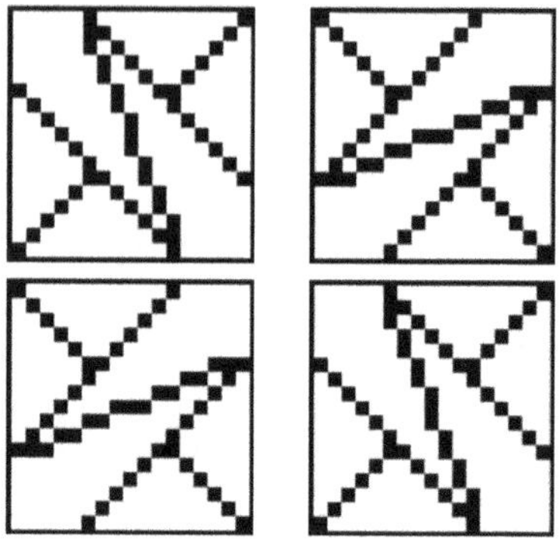

TAAA008 - Rotation

TAAA008SSA

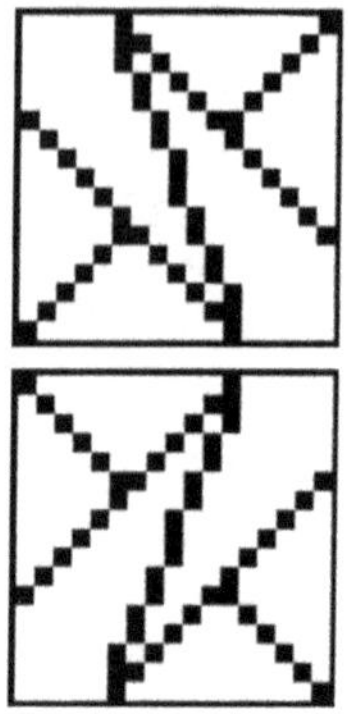

TAAA008 – Horizontal Reflection

TAAA008ASS

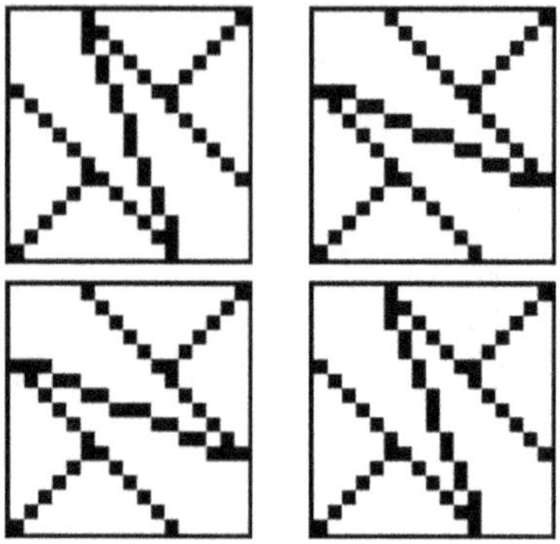

TAAA008 – Rotation and Vertical Reflection

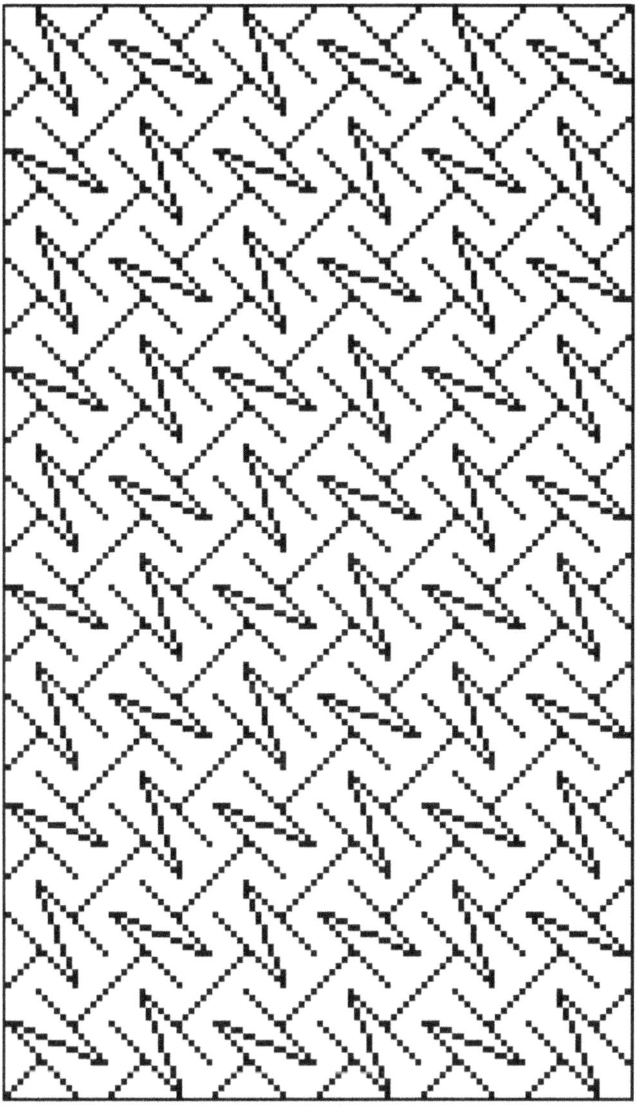

TAAA008SAA

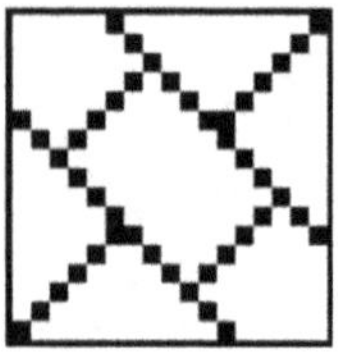

TAAA009

TAAA009SSS

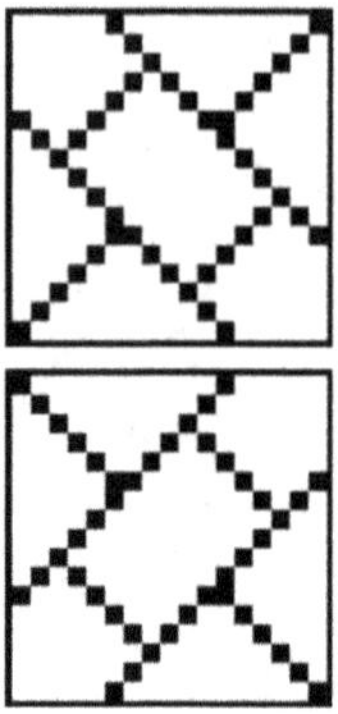

TAAA009 – Horizontal Reflection

TAAA009ASS

Tile Patterns: Symmetry-Based Replication

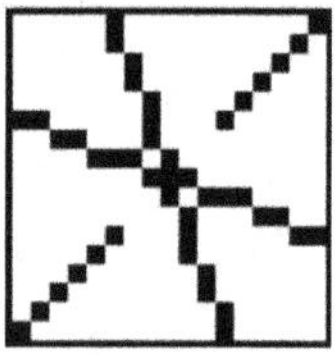

TAAA010

TAAA010SSS

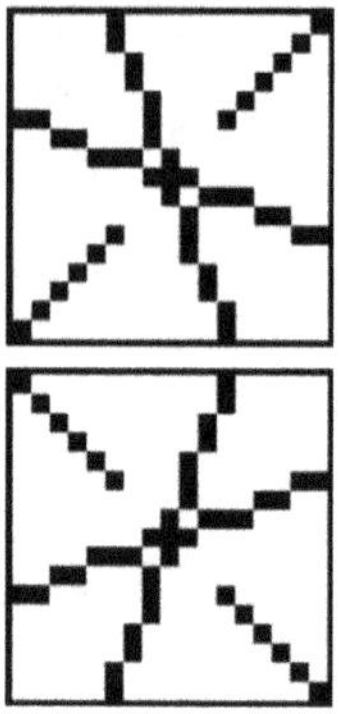

TAAA010 – Horizontal Reflection

TAAA010ASS

Tile Patterns: Symmetry-Based Replication

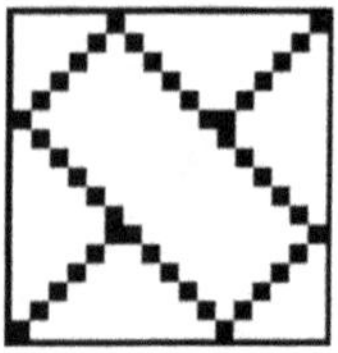

TAAA011

TAAA011SSS

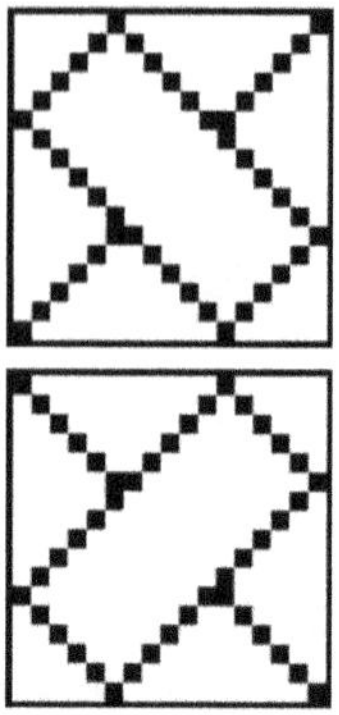

TAAA011

TAAA011ASS

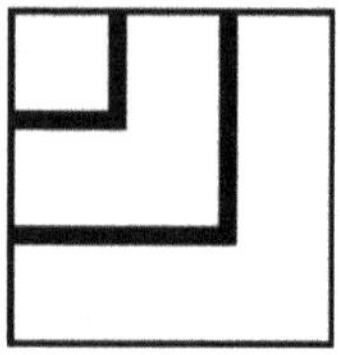

TAAA012

TAAA012SSS

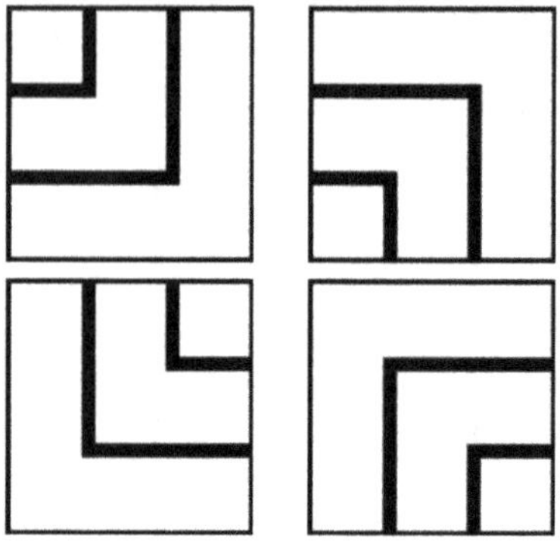

TAAA012 - Rotation

TAAA012SSA

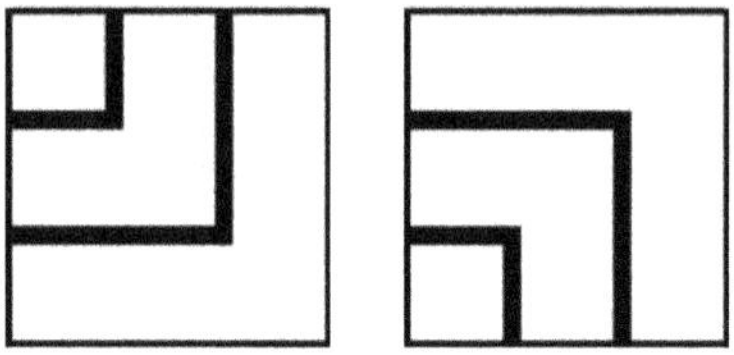

TAAA012 – Vertical Reflection

TAAA012SAS

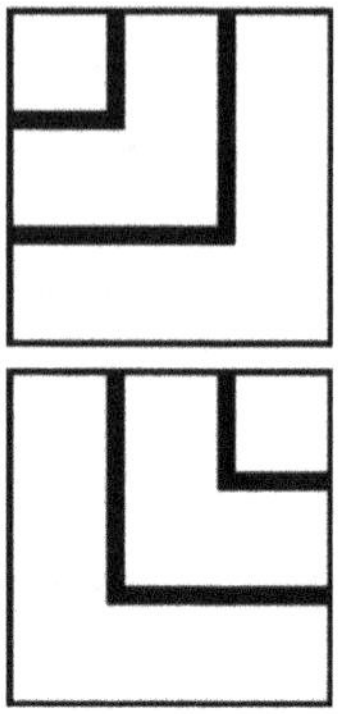

TAAA012 – Horizontal Reflection

TAAA012ASS

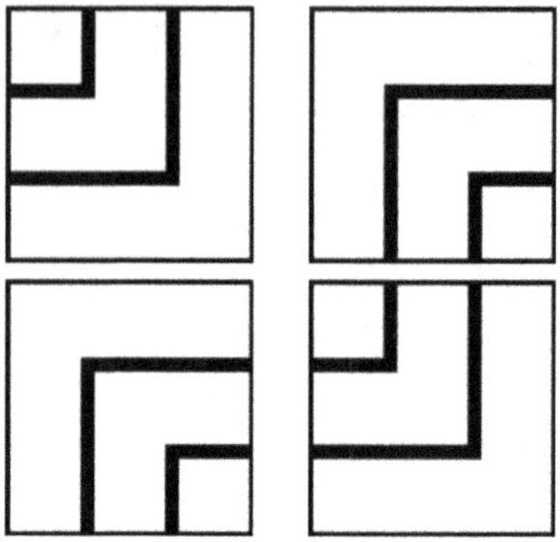

TAAA012 – Rotation and Vertical Reflection

TAAA012SAA

TAAA012 – Vertical and Horizontal Reflections

TAAA012AAS

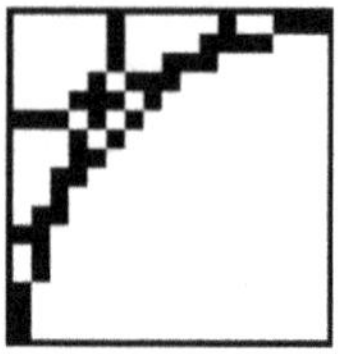

TAAA013

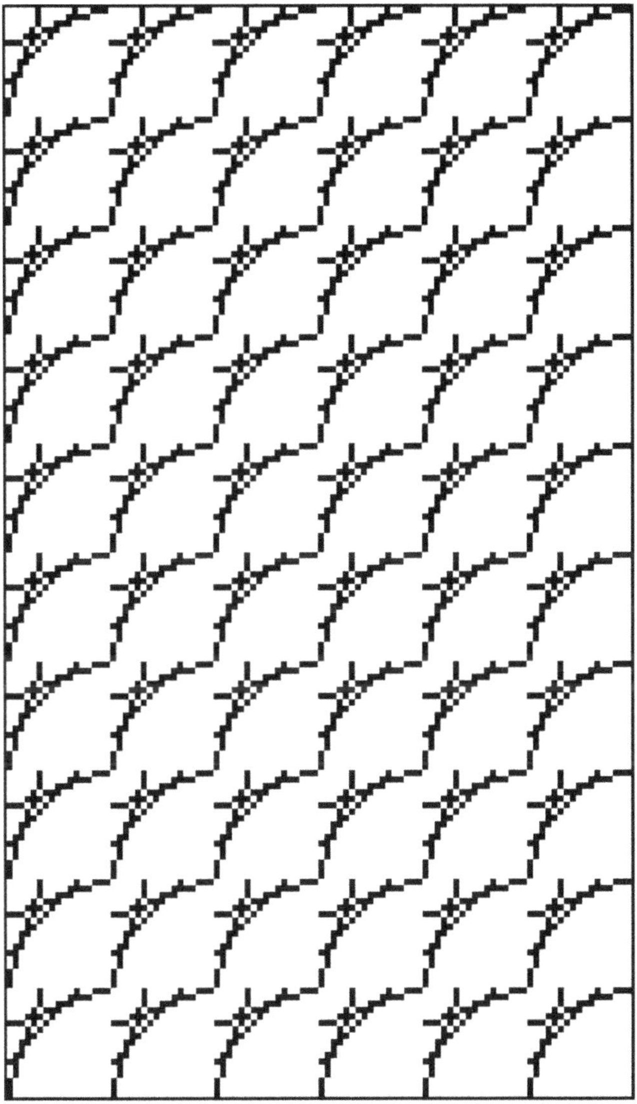

TAAA013SSS

Tile Patterns: Symmetry-Based Replication

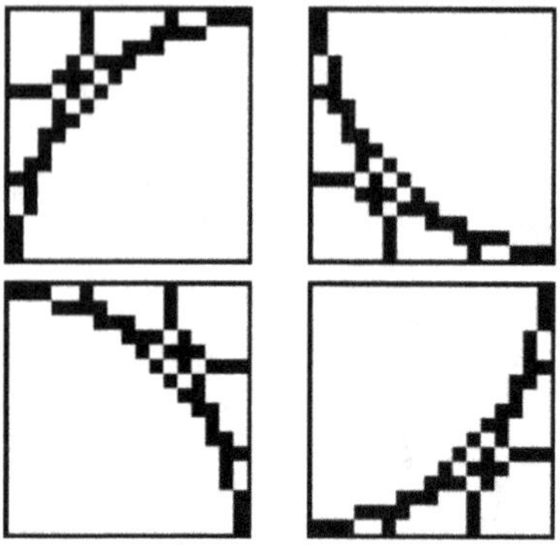

TAAA013

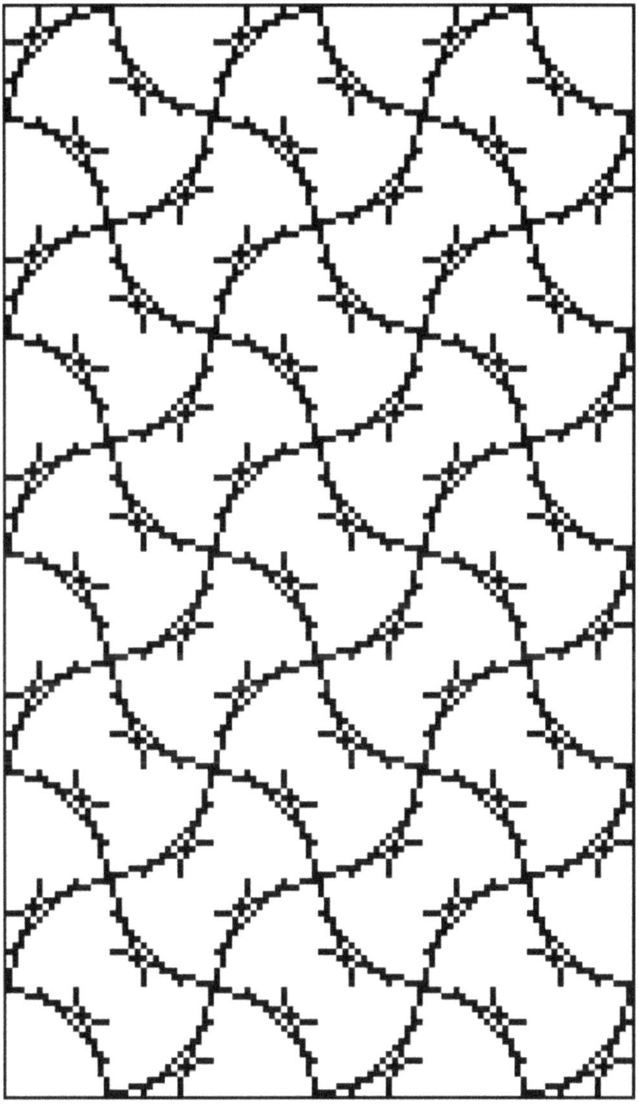

TAAA013SSA

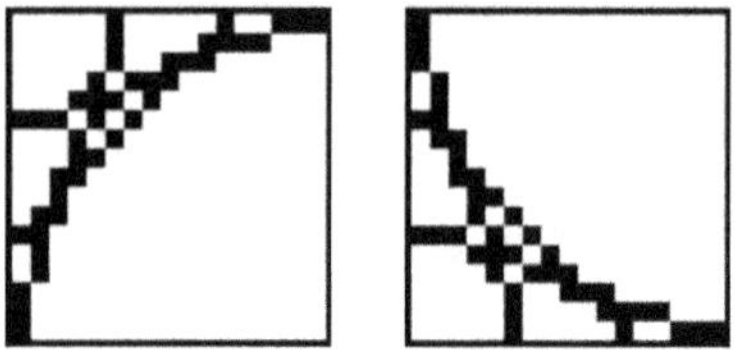

TAAA013 – Vertical Reflection

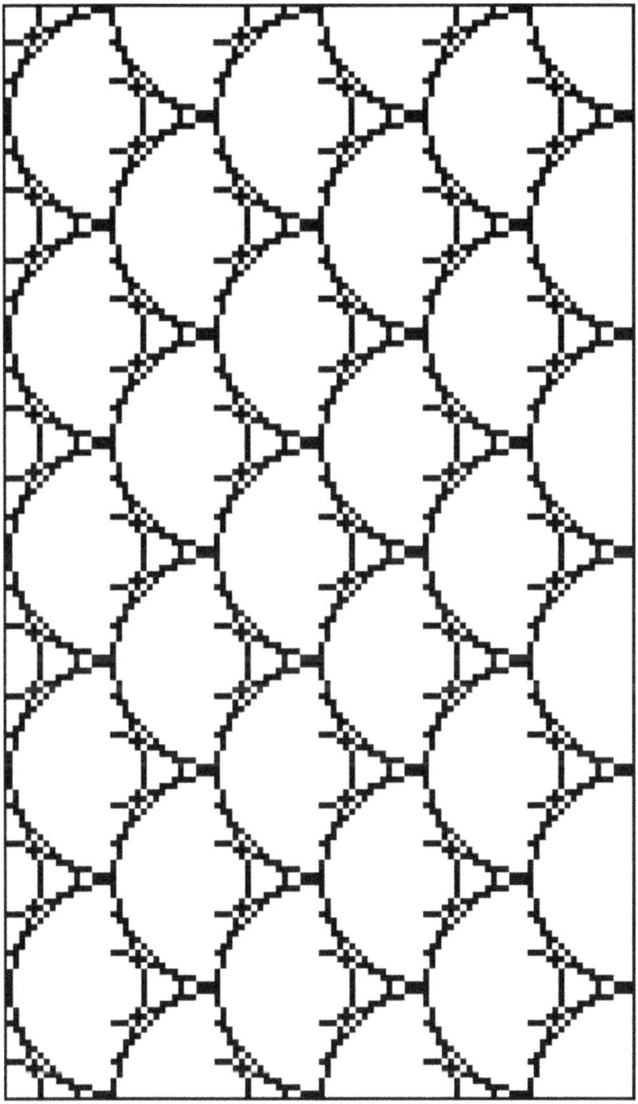

TAAA013SAS

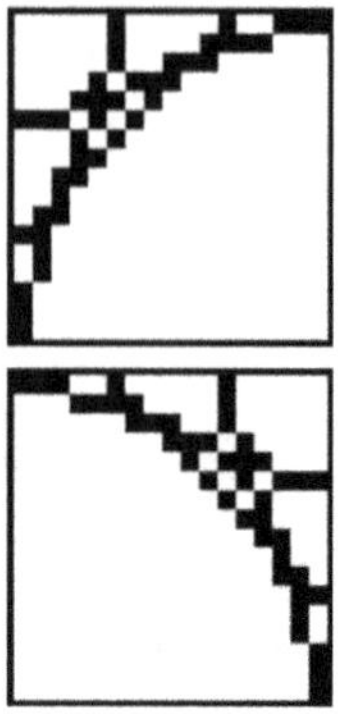

TAAA013 – Horizontal Reflection

TAAA013ASS

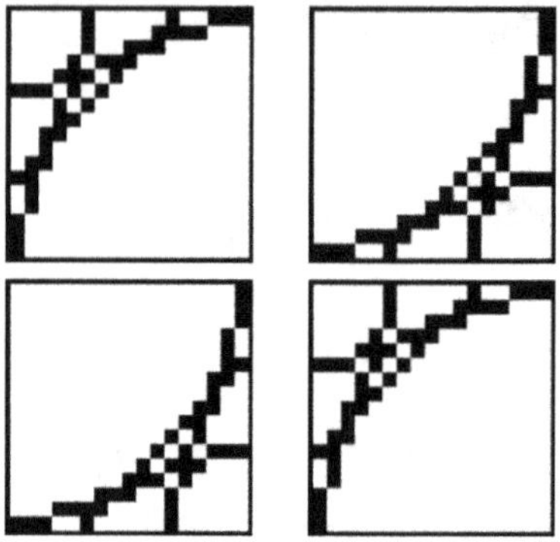

TAAA013 – Rotation and
Vertical Reflection

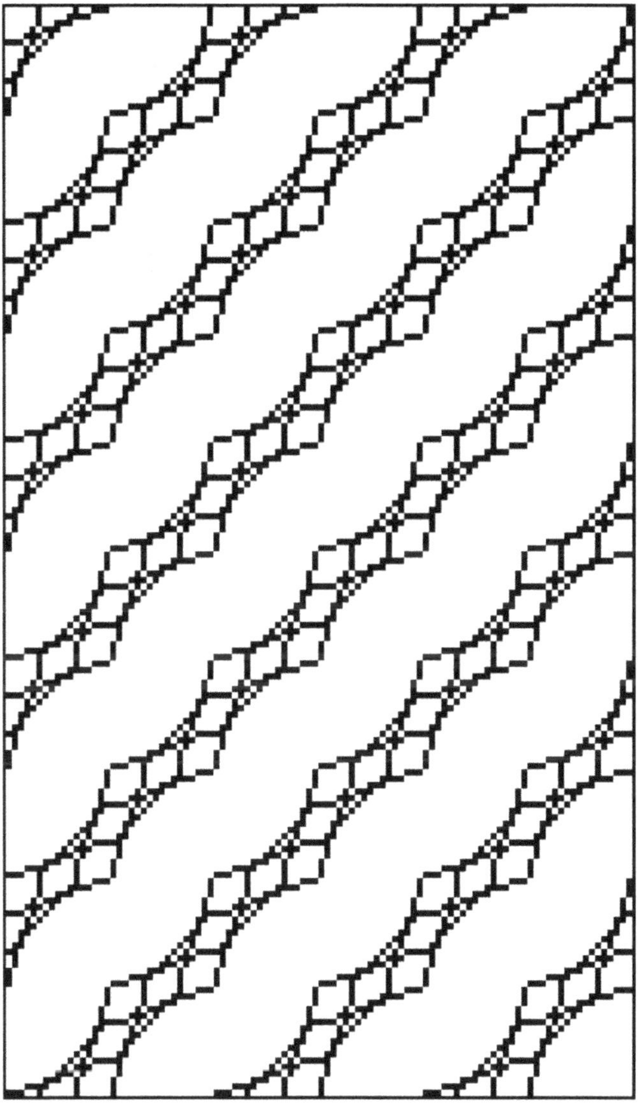

TAAA013SAA

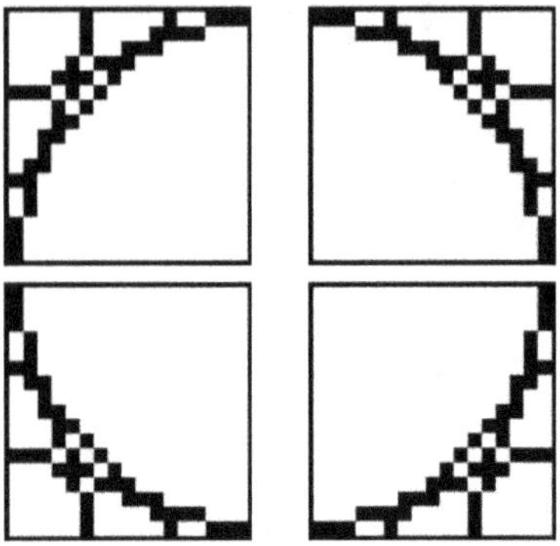

TAAA013 – Vertical and Horizontal Reflections

TAAA013AAS

TAAA014

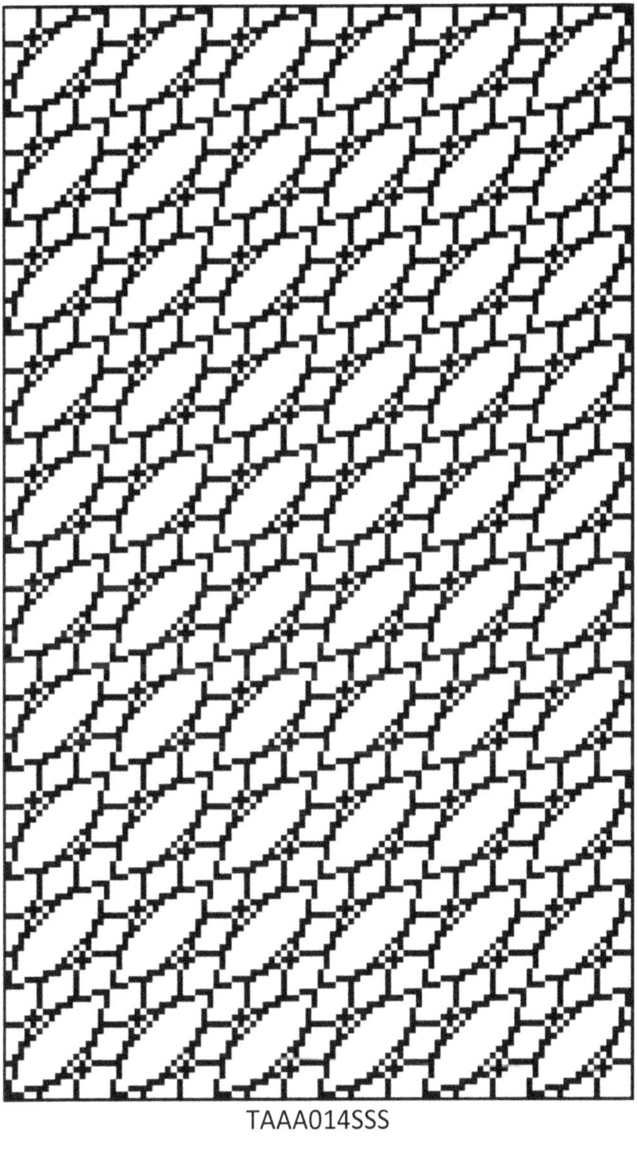

TAAA014SSS

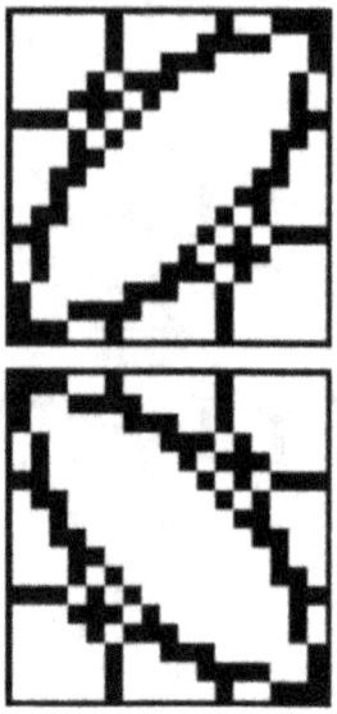

TAAA014 – Horizontal Reflection

TAAA014ASS

Tile Patterns: Symmetry-Based Replication

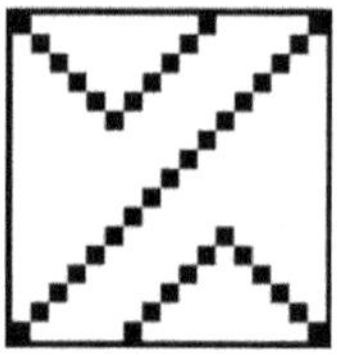

TAAA015

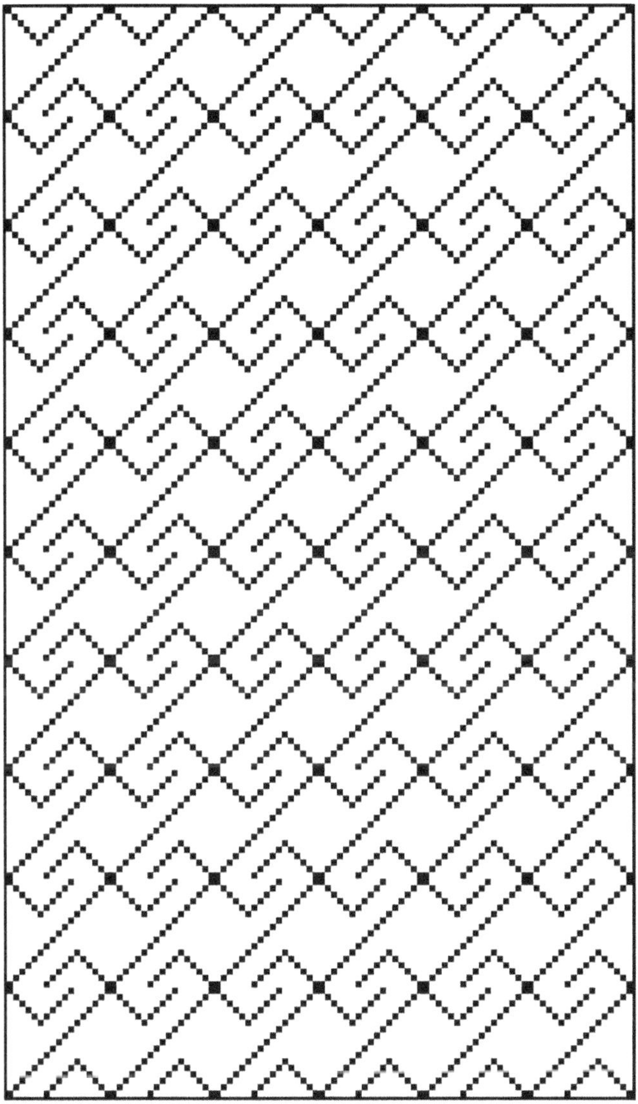

TAAA015SSS

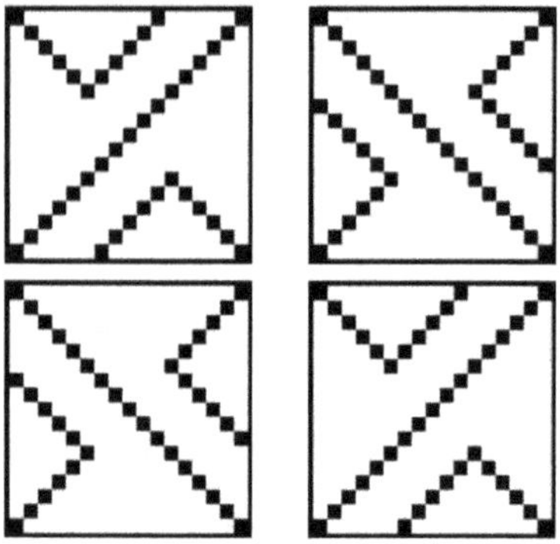

TAAA015 - Rotation

TAAA015SSA

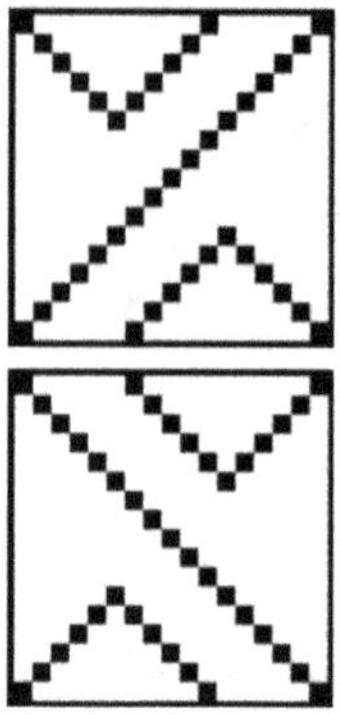

TAAA015 – Horizontal Reflection

TAAA015ASS

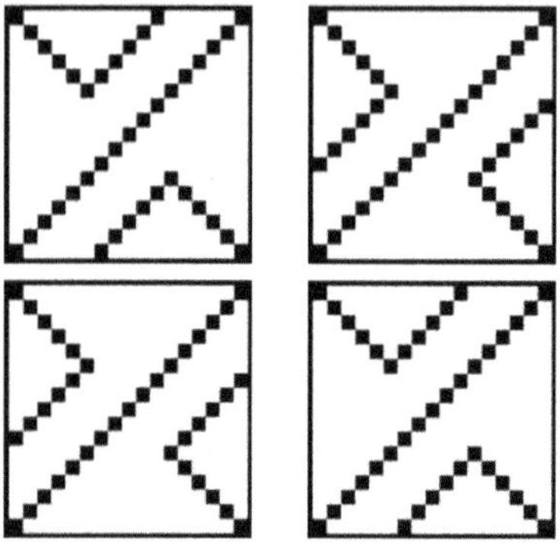

TAAA015 – Rotation and Vertical Reflection

TAAA015SAA

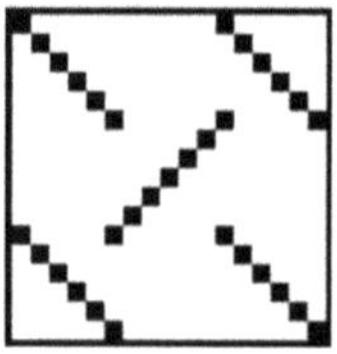

TAAA016

TAAA016SSS

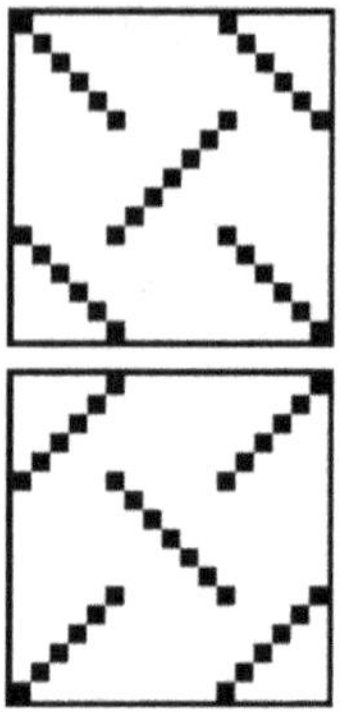

TAAA016 – Horizontal Reflection

TAAA016ASS

TAAA017

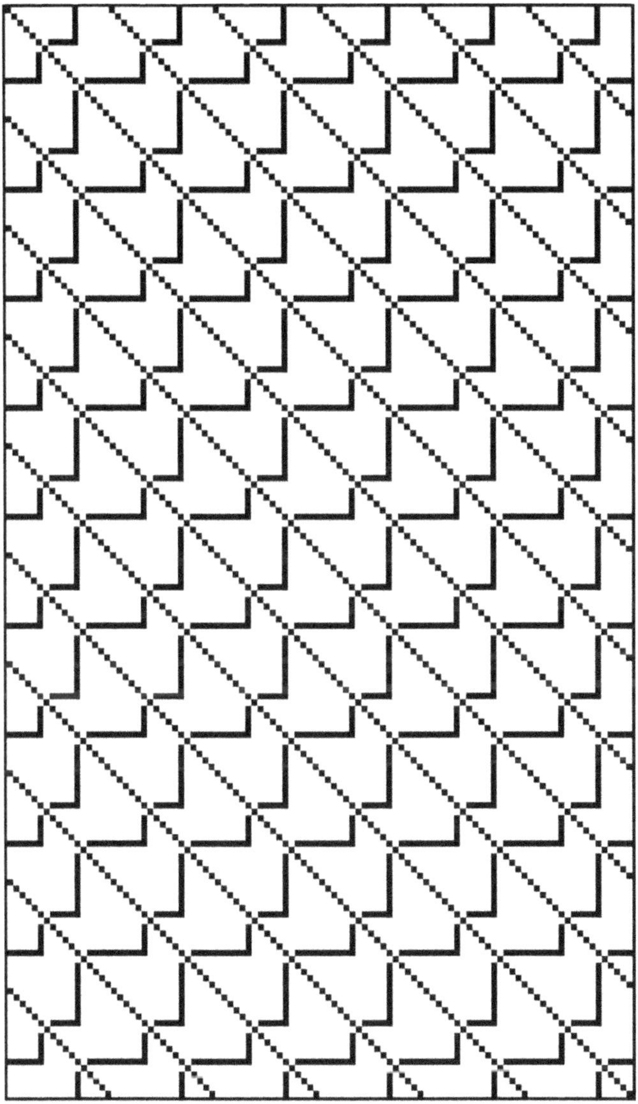

TAAA017SSS

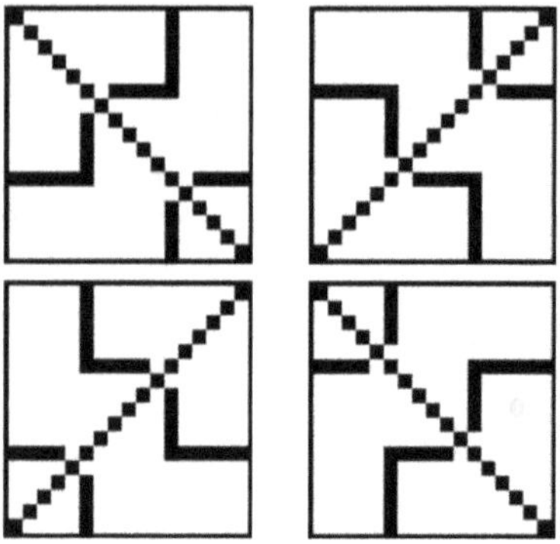

TAAA017 - Rotation

TAAA017SSA

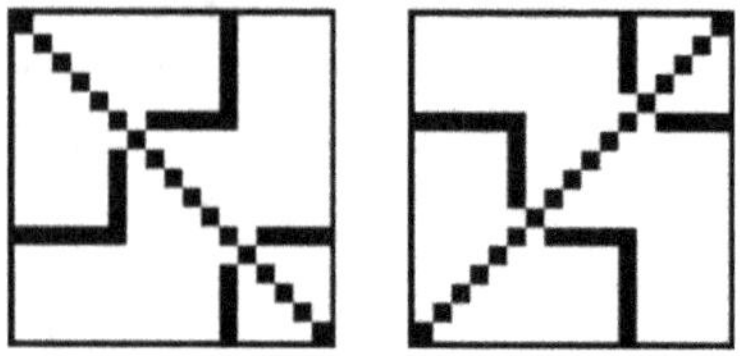

TAAA017 – Vertical Reflection

TAAA017SAS

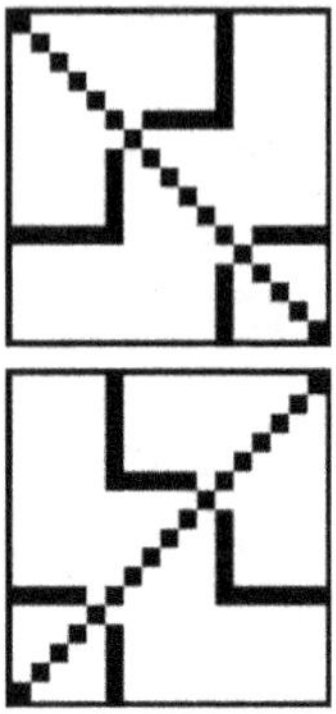

TAAA017 – Horizontal Reflection

TAAA017ASS

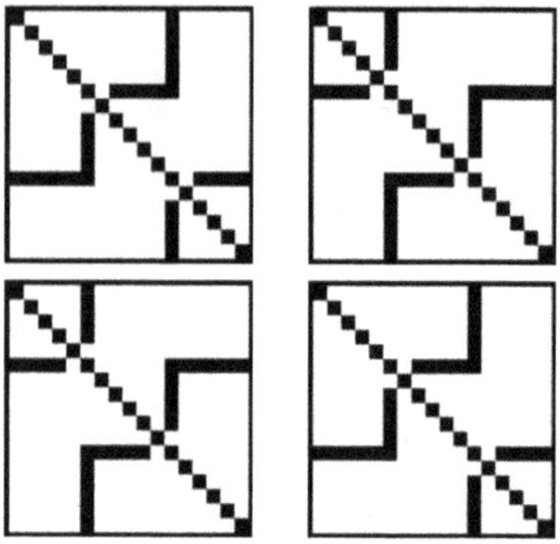

TAAA017 – Rotation and Vertical Reflection

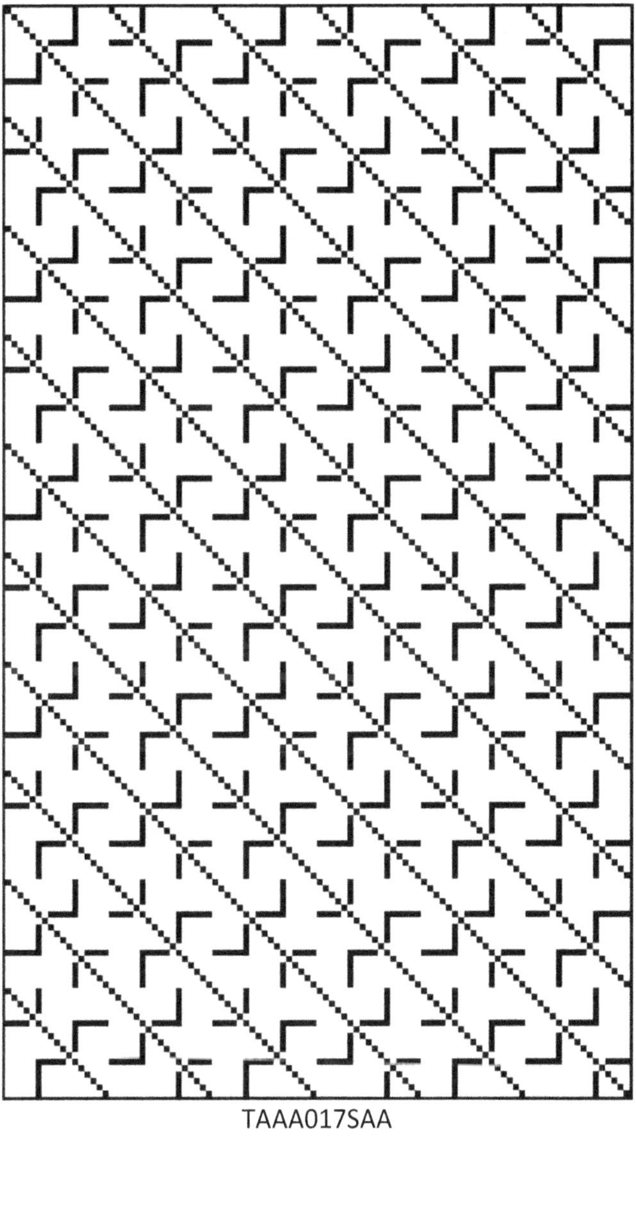

TAAA017SAA

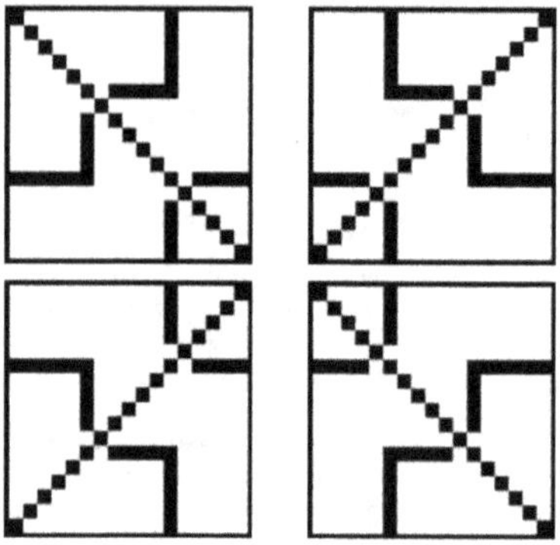

TAAA017 – Vertical and Horizontal Reflections

TAAA017AAS

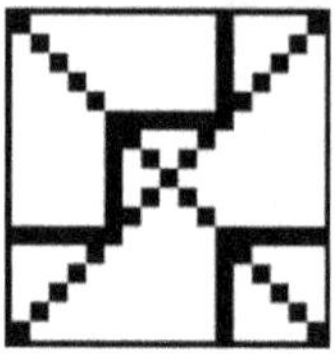

TAAA018

TAAA018SSS

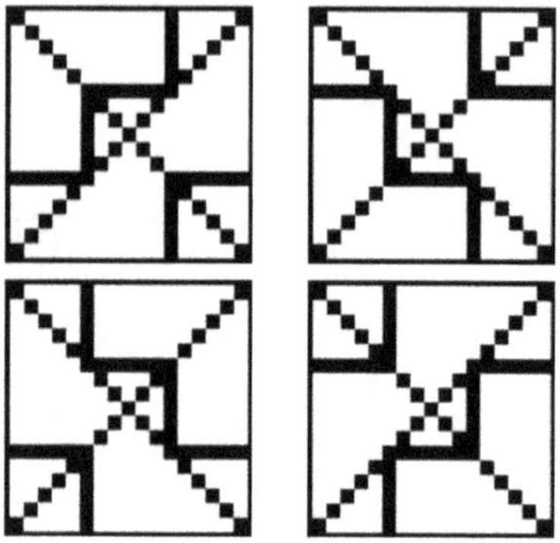

TAAA018 - Rotation

TAAA018SSA

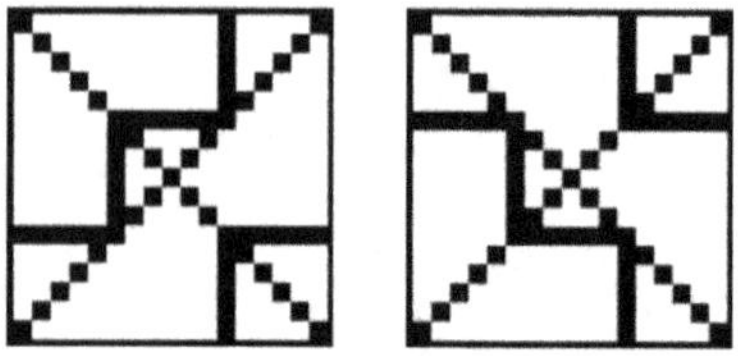

TAAA018 – Vertical Reflection

TAAA018SAS

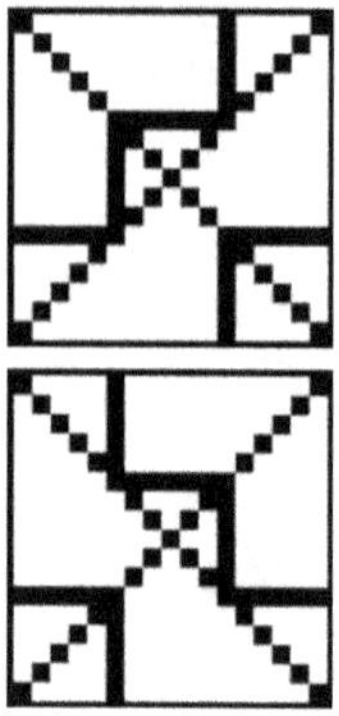

TAAA018 – Horizontal Reflection

TAAA018ASS

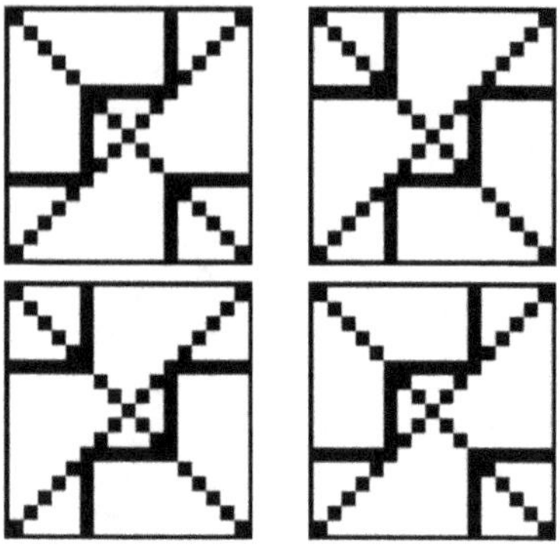

TAAA018 – Rotation and Vertical Reflection

TAAA018SAA

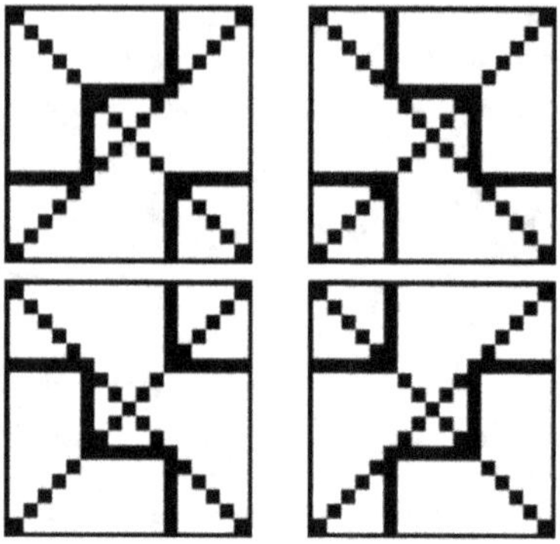

TAAA018 – Vertical and Horizontal Reflections

TAAA018AAS

Tile Patterns: Symmetry-Based Replication

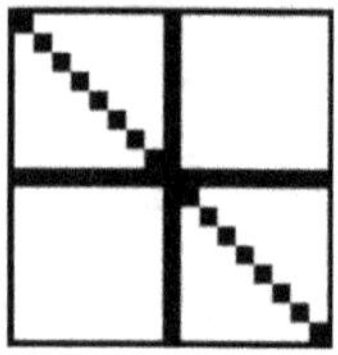

TAAA019

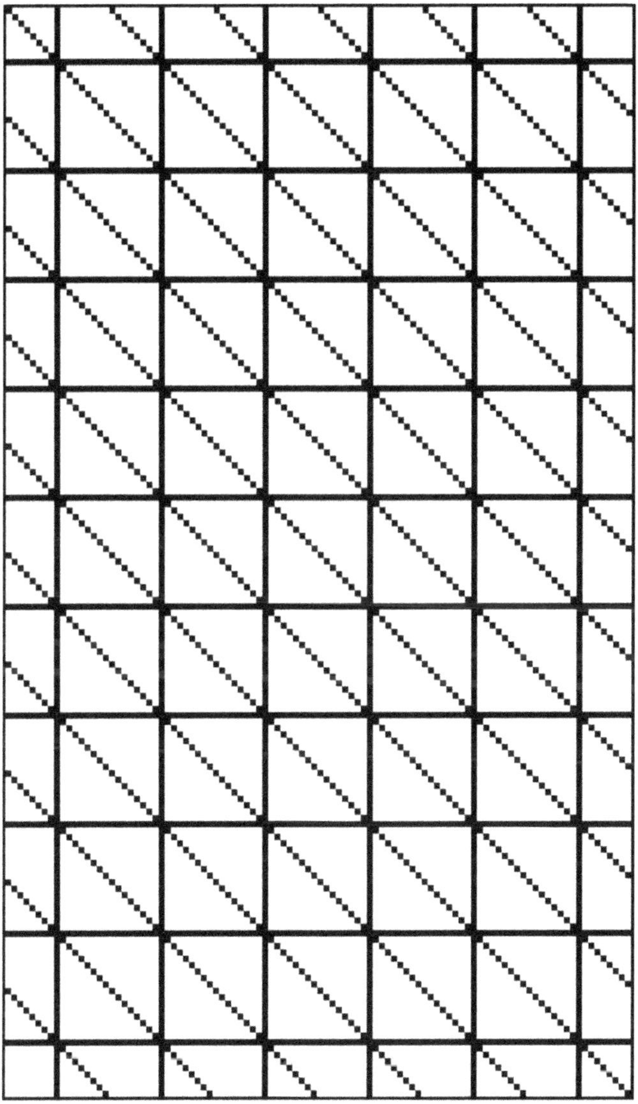

TAAA019SSS

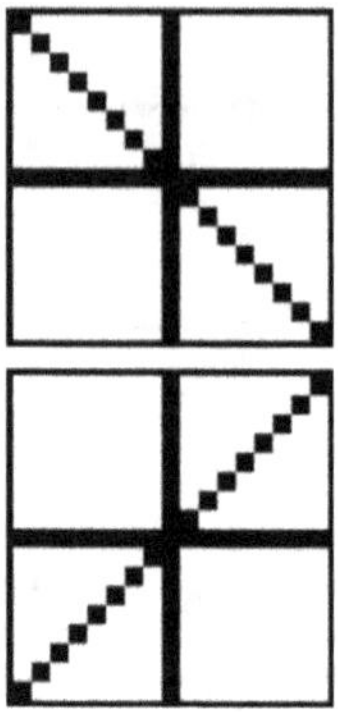

TAAA019 – Horizontal Reflection

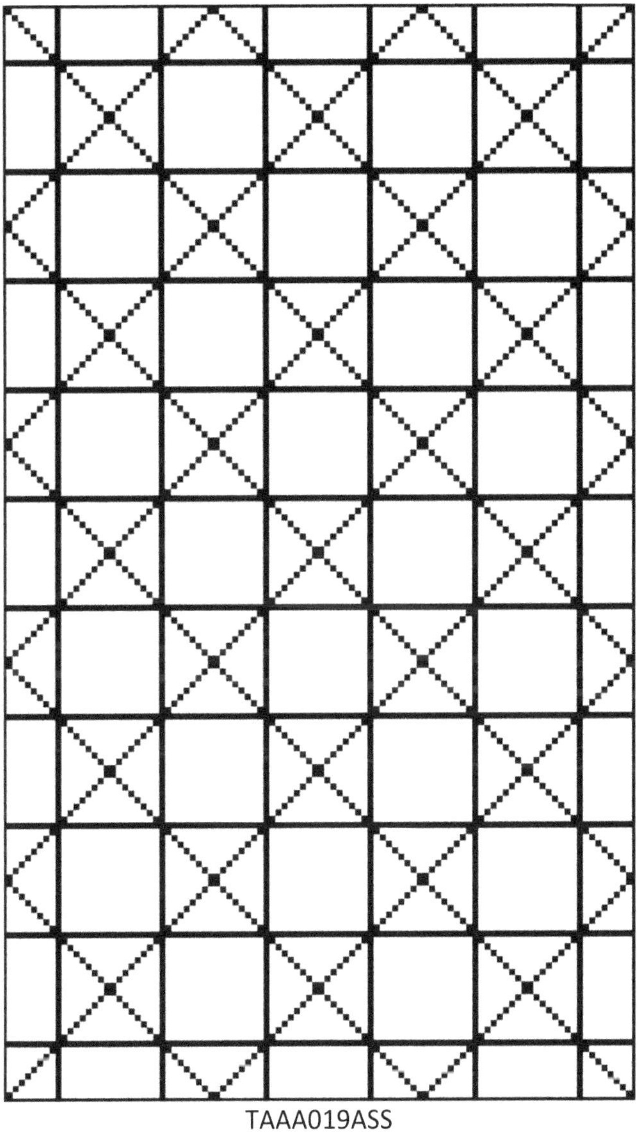

TAAA019ASS

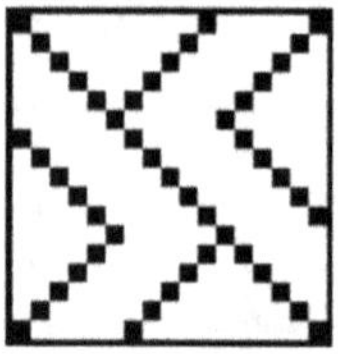

TAAA020

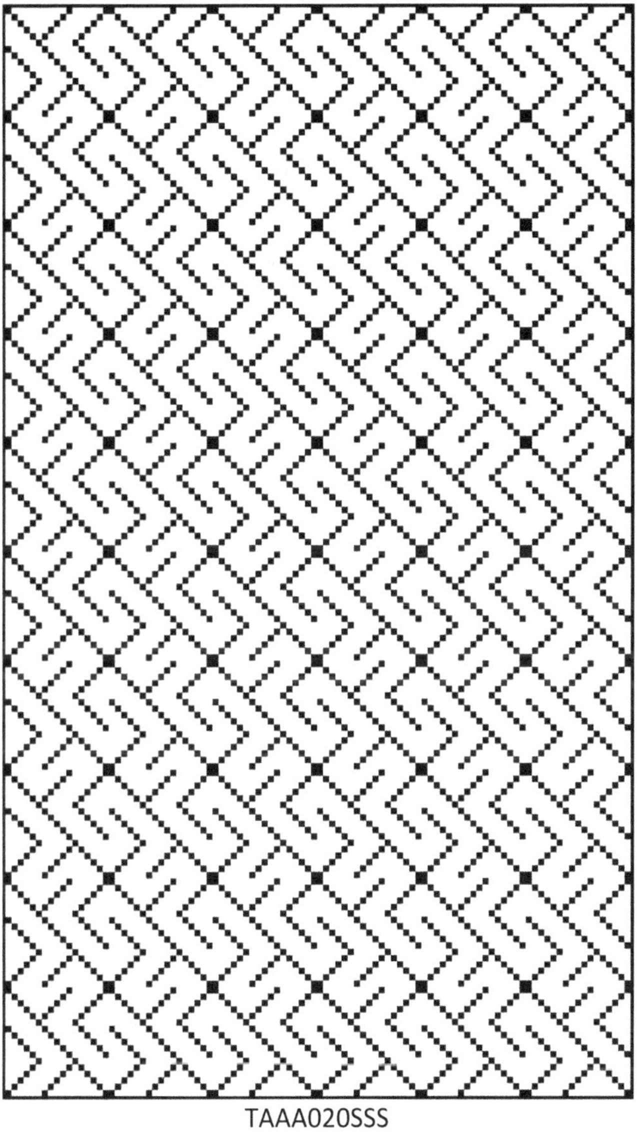

TAAA020SSS

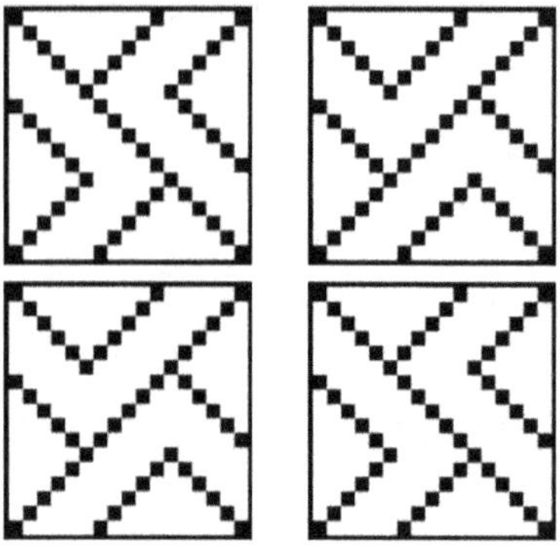

TAAA020 - Rotation

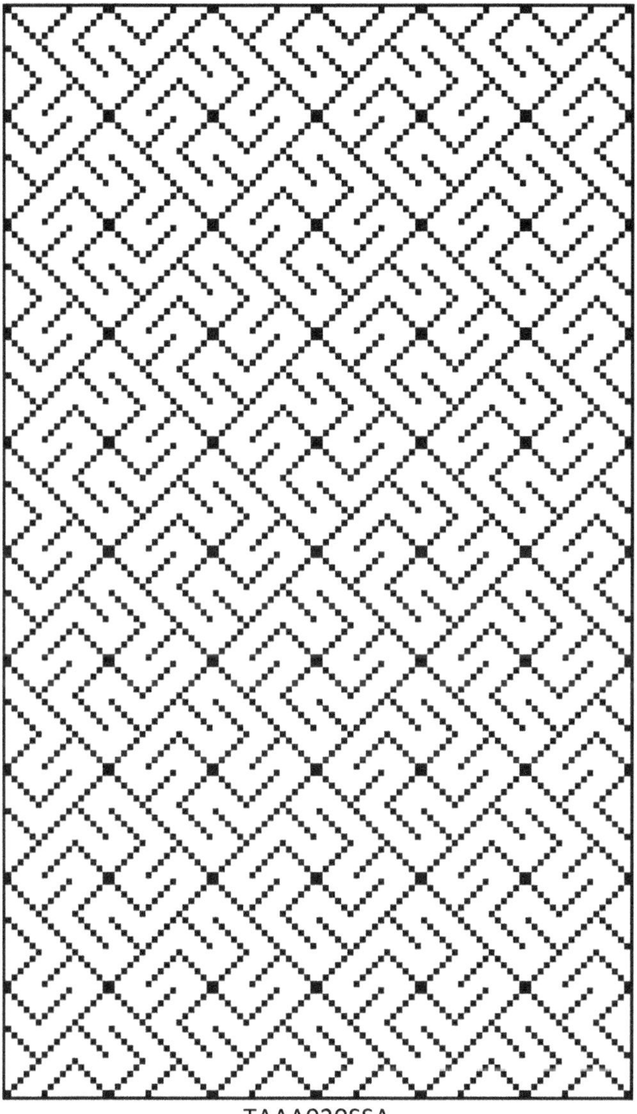

TAAA020SSA

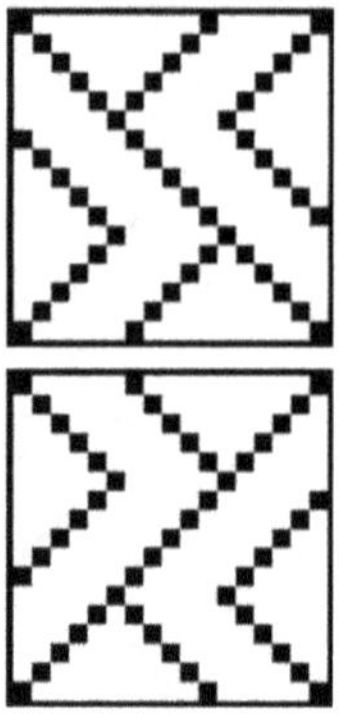

TAAA020 – Horizontal Reflection

TAAA020ASS

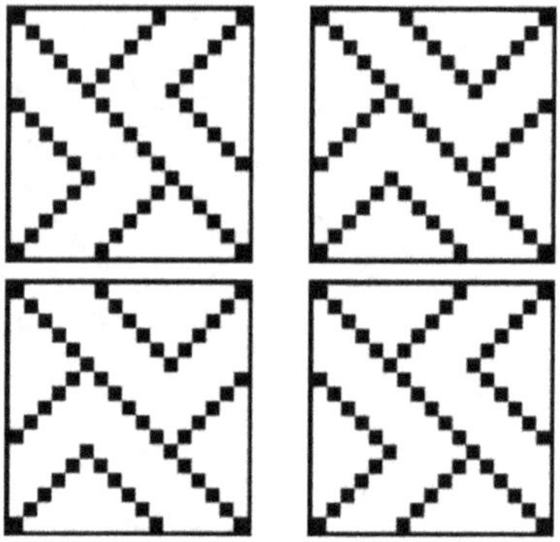

TAAA020 – Rotation and Vertical Reflection

TAAA020SAA

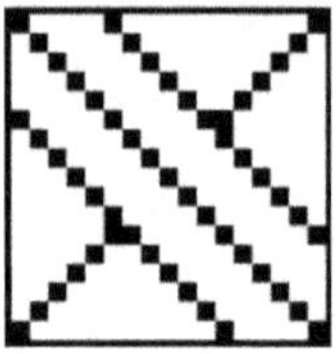

TAAA021

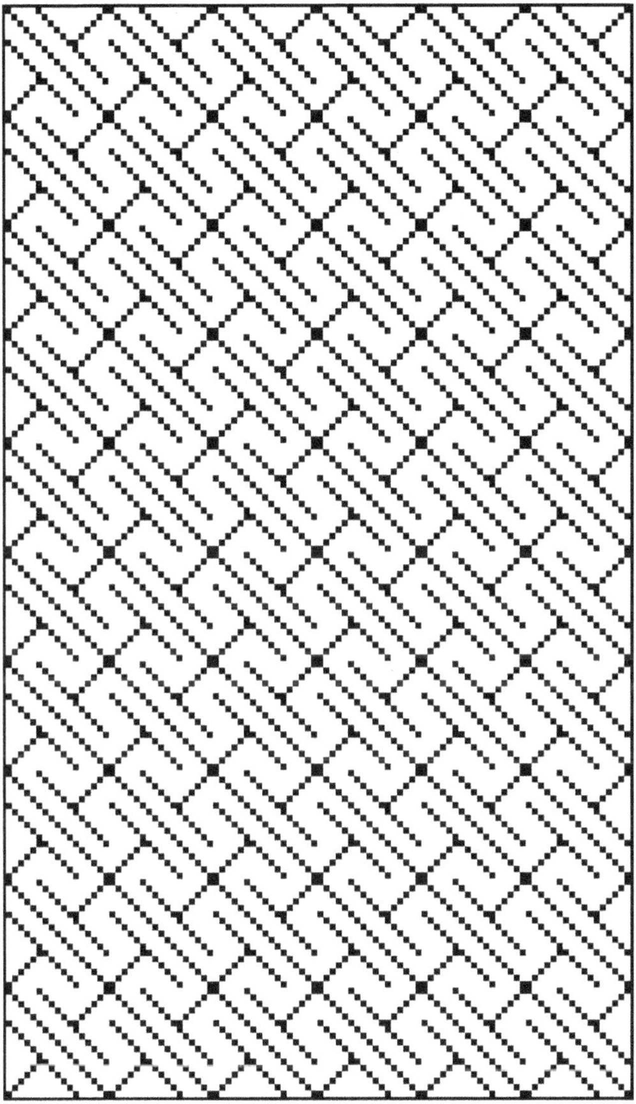

TAAA021SSS

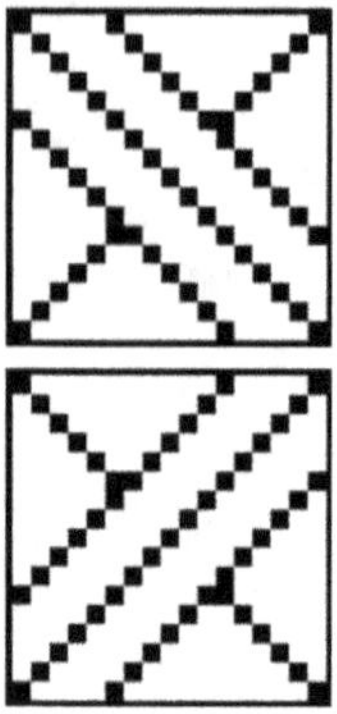

TAAA021N – Horizontal Reflection

TAAA021ASS

TAAA022

TAAA022SSS

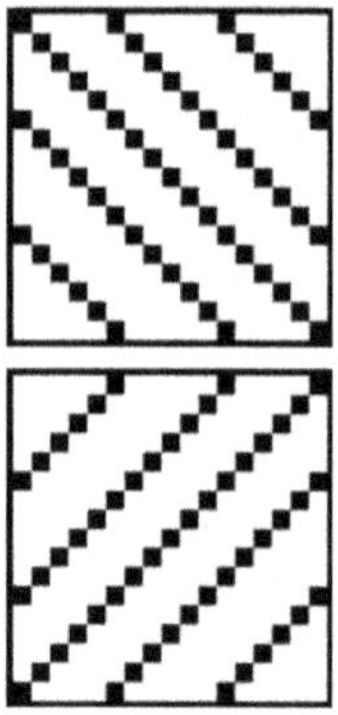

TAAA022 – Horizontal Reflection

TAAA022ASS

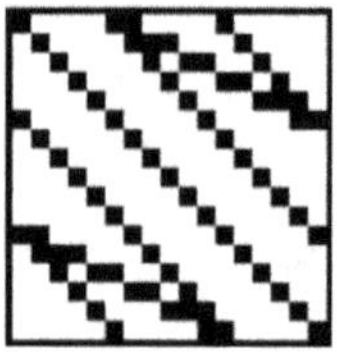

TAAA023

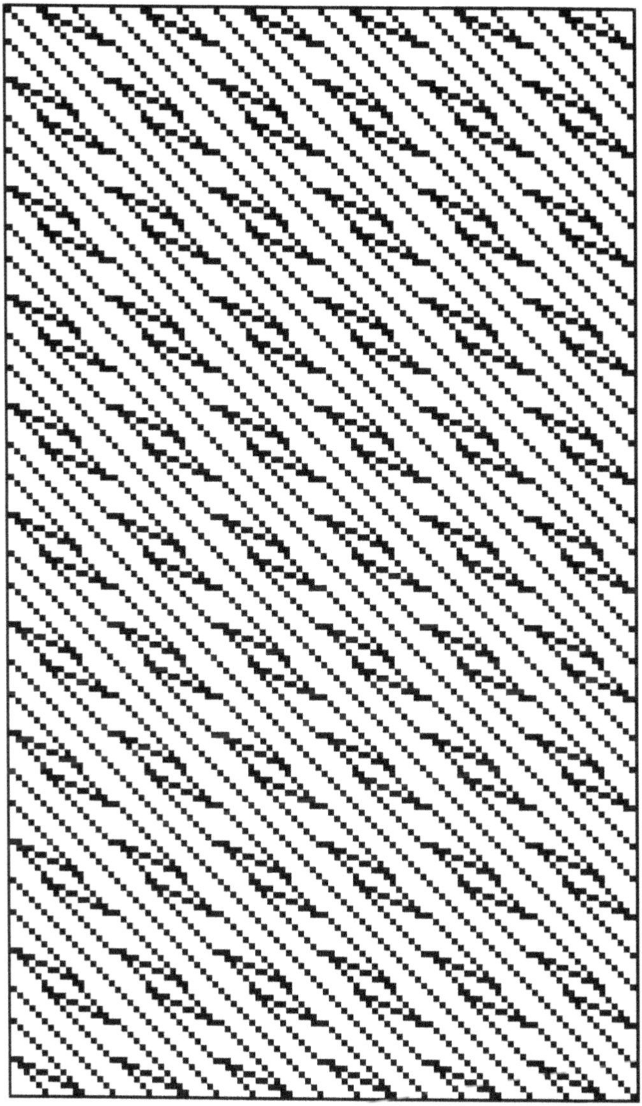

TAAA023SSS

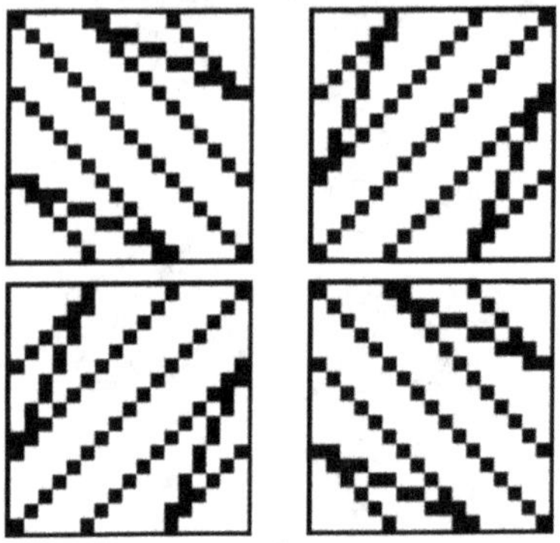

TAAA023 - Rotation

TAAA023SSA

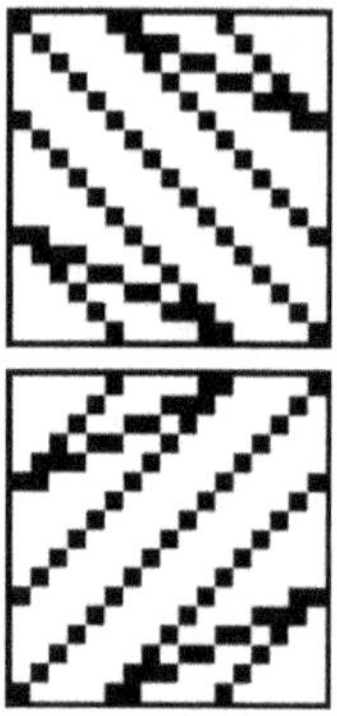

TAAA023 – Horizontal Reflection

TAAA023ASS

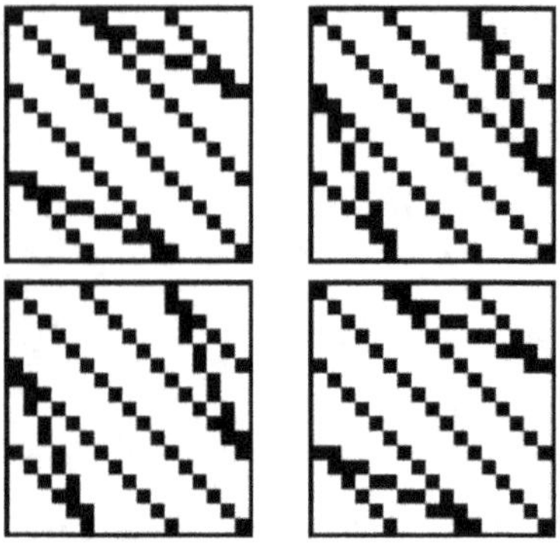

TAAA023 – Rotation and Vertical Reflection

TAAA023SAA

Tile Patterns: Symmetry-Based Replication

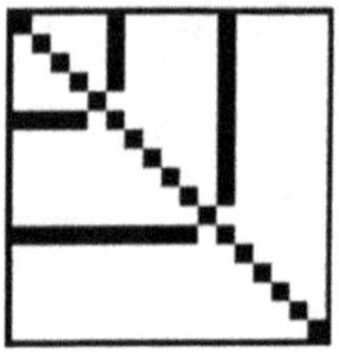

TAAA024

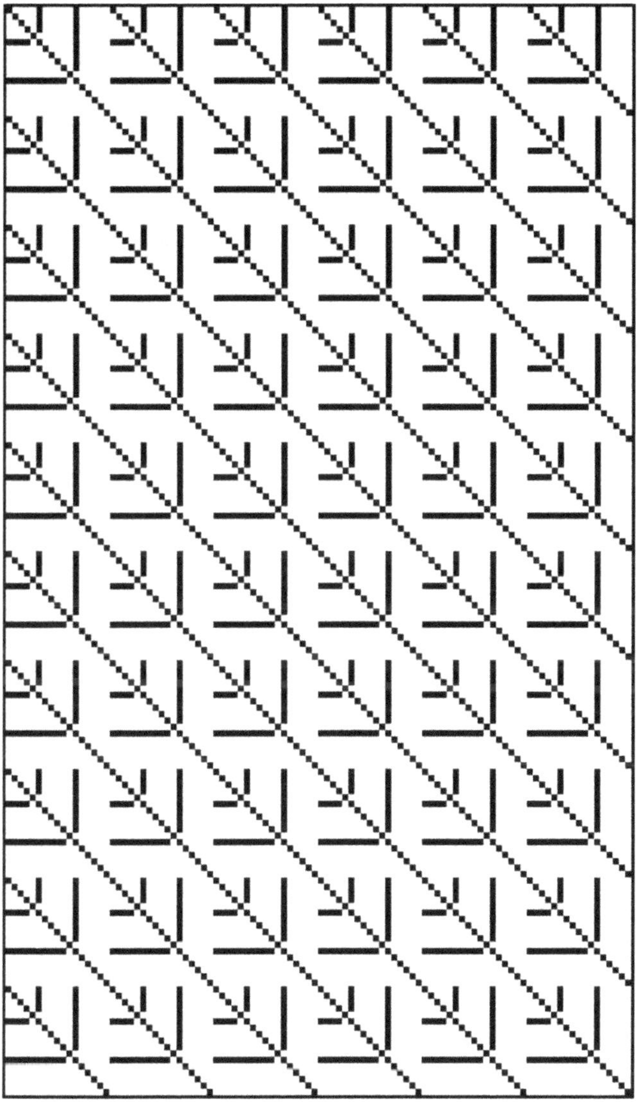

TAAA024SSS

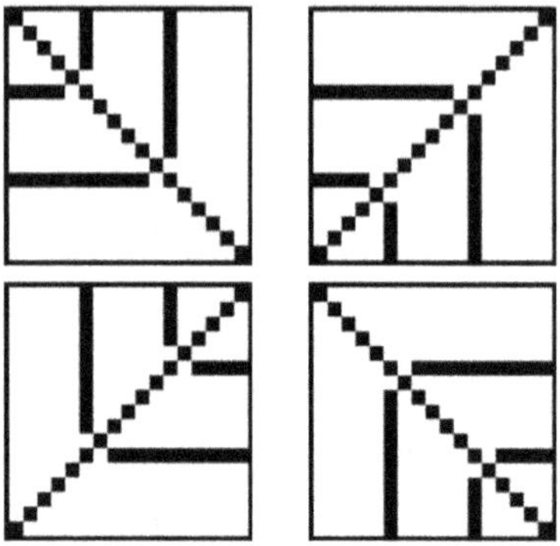

TAAA024 - Rotation

TAAA024SSA

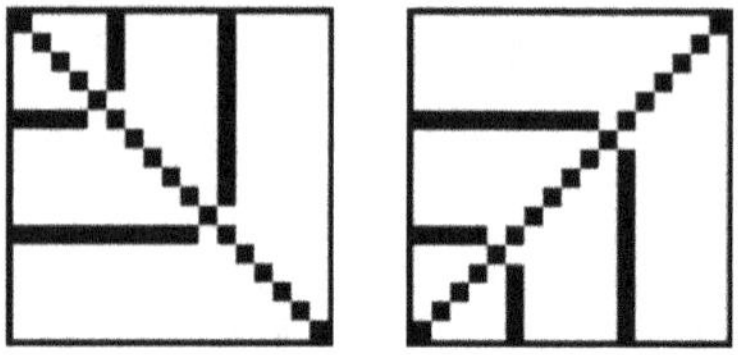

TAAA024 – Vertical Reflection

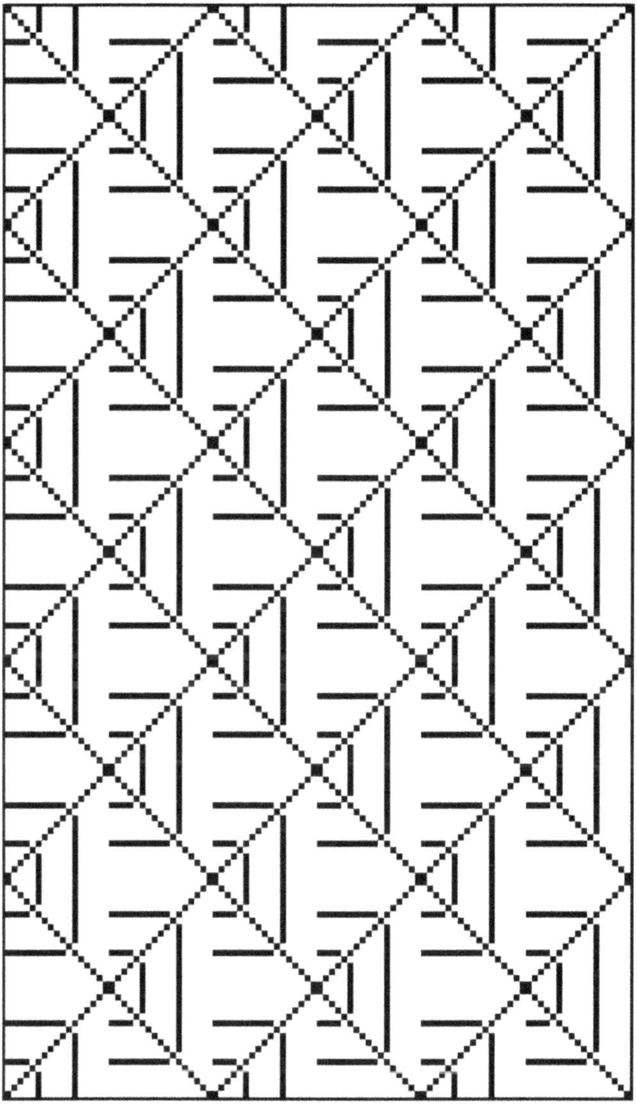

TAAA024SAS

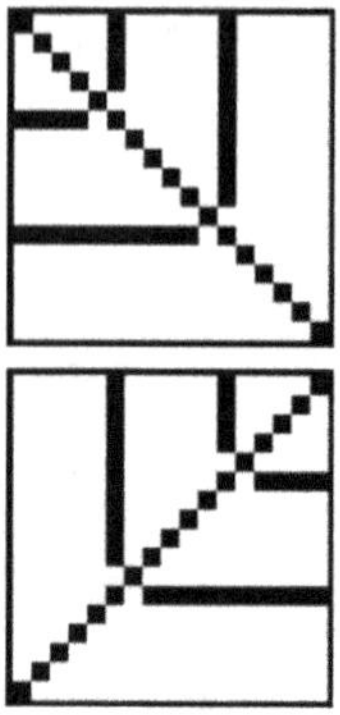

TAAA024 – Horizontal Reflection

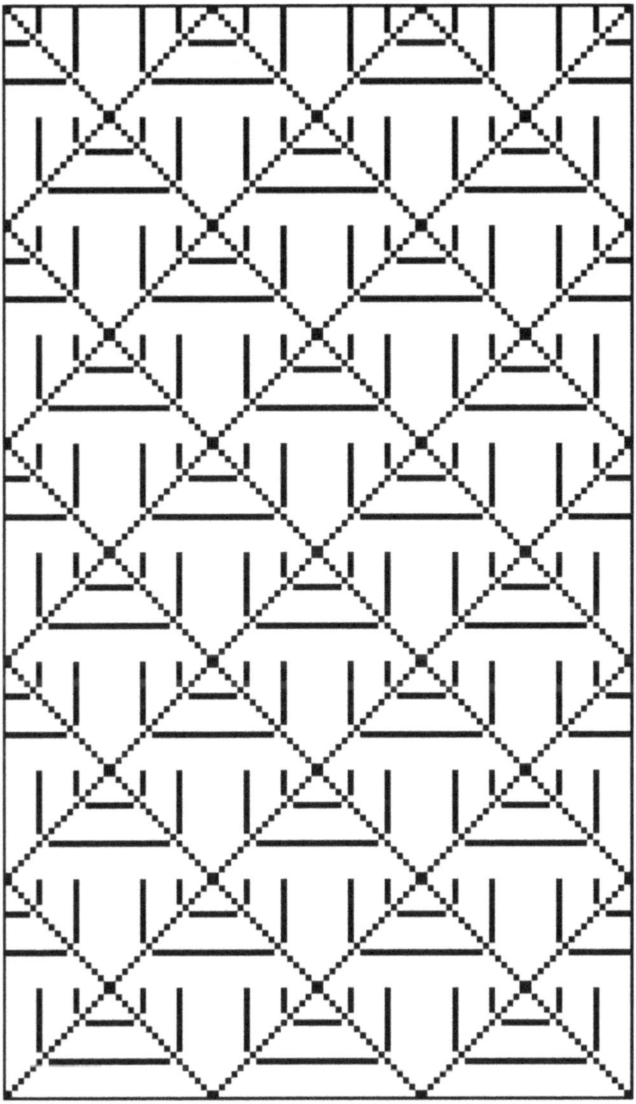

TAAA024ASS

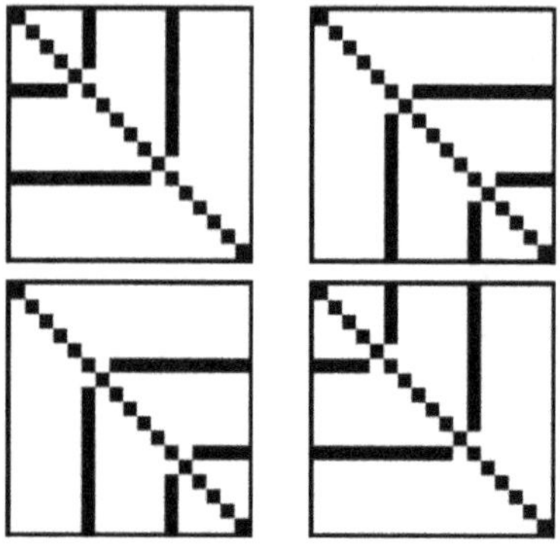

TAAA024 – Rotation and Vertical Reflection

TAAA024SAA

TAAA024 – Vertical and Horizontal Reflections

TAAA024AAS

Tile Patterns: Symmetry-Based Replication

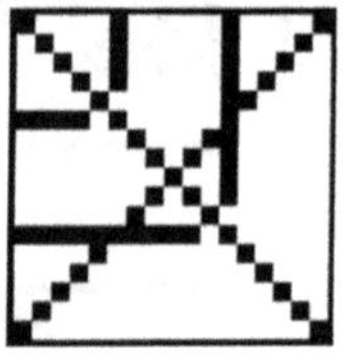

TAAA025

TAAA025SSS

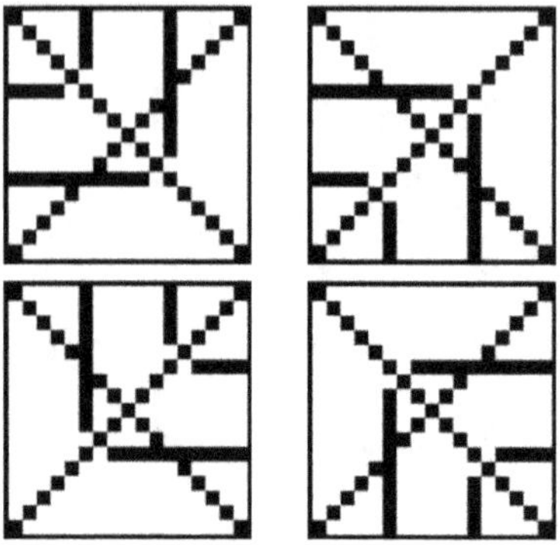

TAAA025 - Rotation

TAAA025SSA

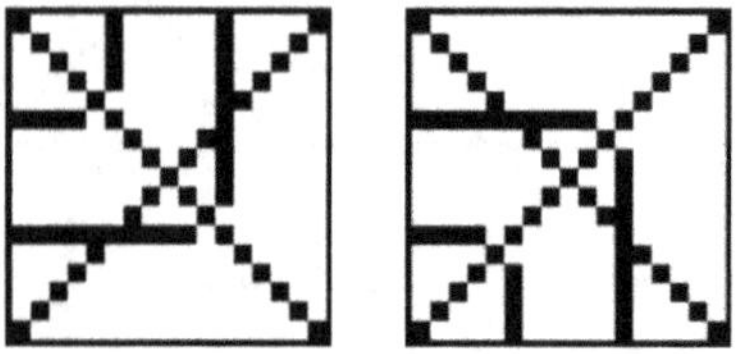

TAAA025 – Vertical Reflection

TAAA025SAS

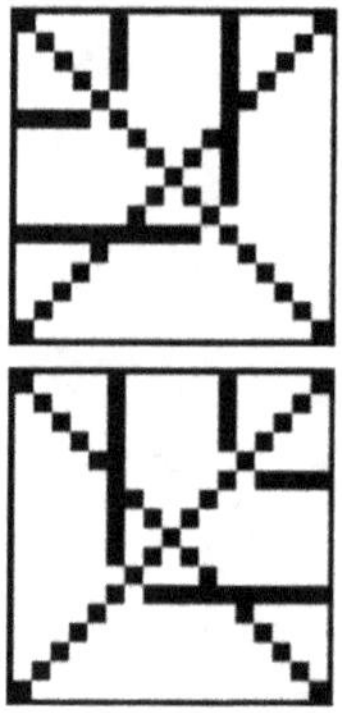

TAAA025 – Horizontal Reflection

TAAA025ASS

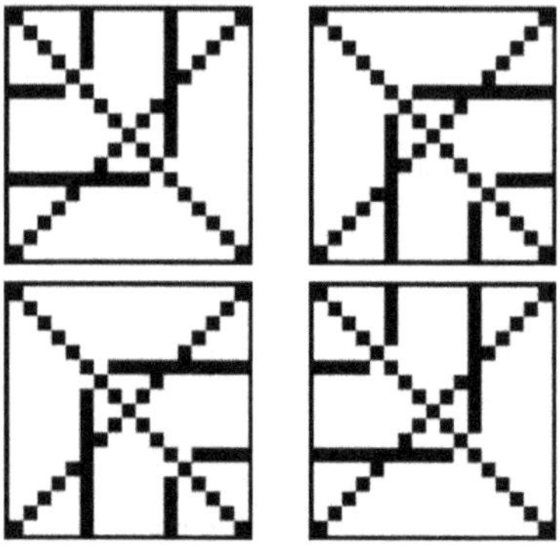

TAAA025 – Rotation and
Vertical Reflection

TAAA025SAA

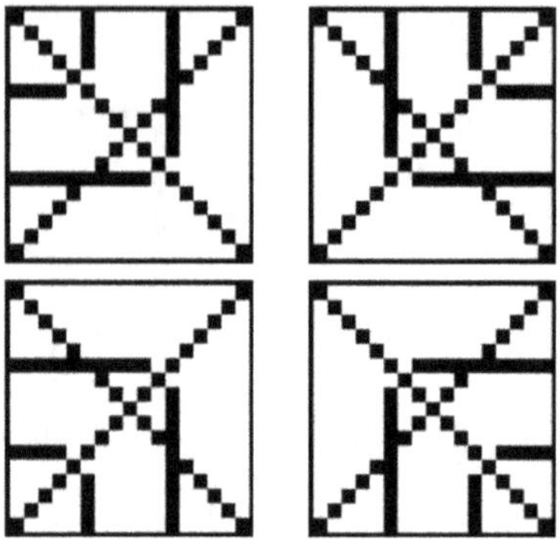

TAAA025 – Vertical and Horizontal Reflections

TAAA025AAS

TAAA026

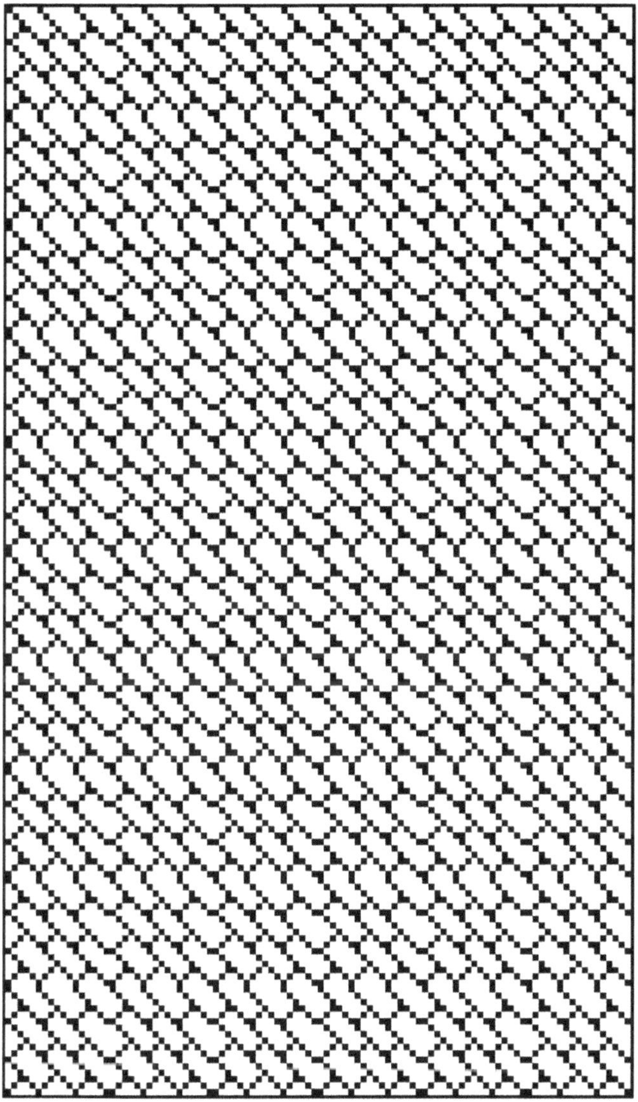

TAAA026SSS

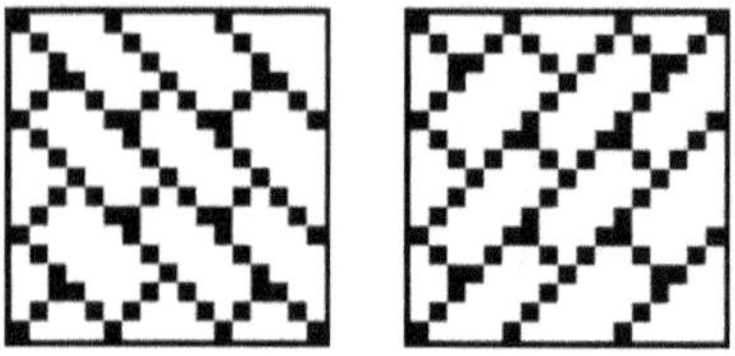

TAAA026 – Vertical Reflection

TAAA026SAS

TAAA026 – Horizontal Reflection

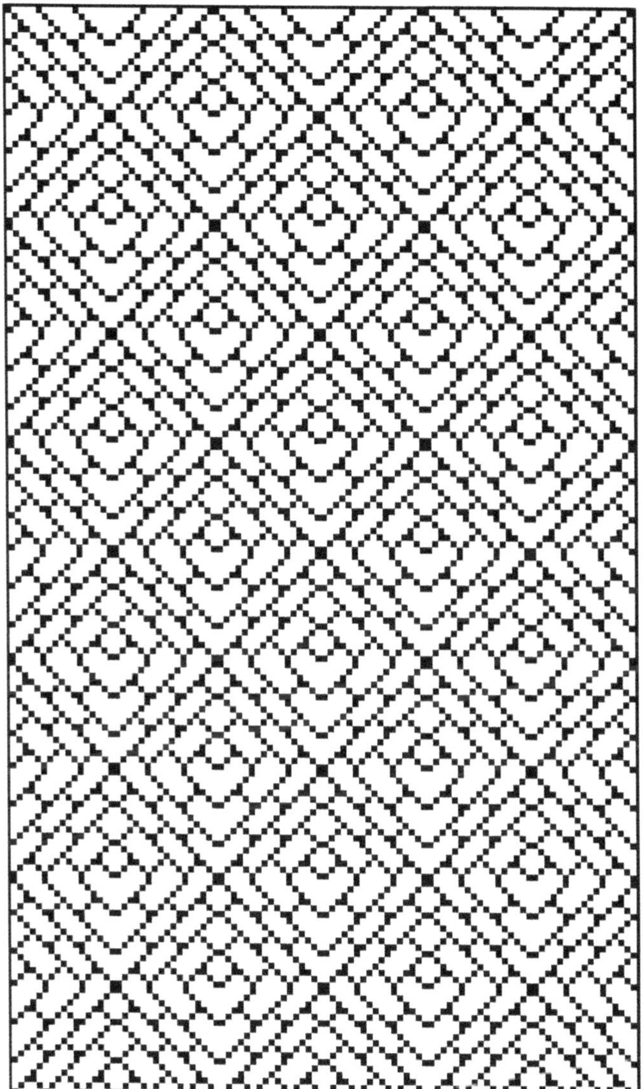

TAAA026ASS

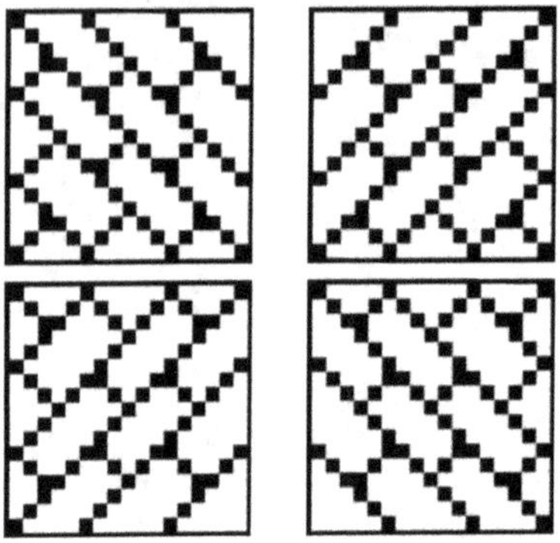

TAAA026 – Vertical and Horizontal Reflections

TAAA026AAS

TAAA027

TAAA027SSS

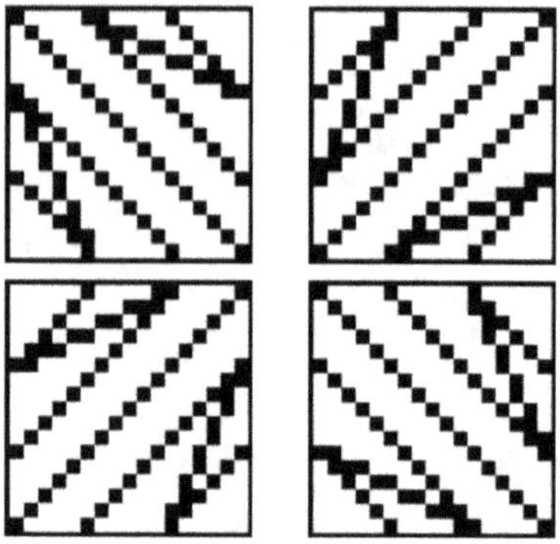

TAAA027 - Rotation

TAAA027SSA

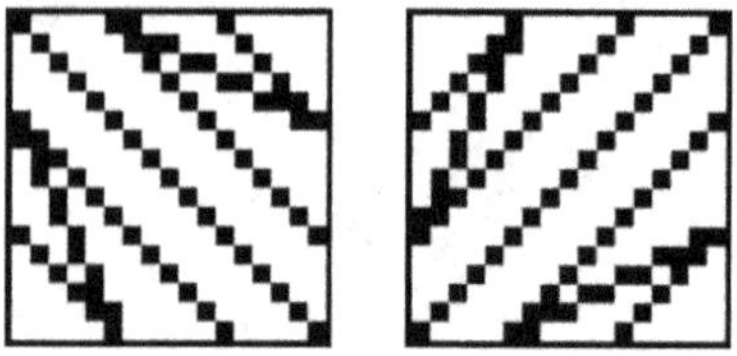

TAAA027 – Vertical Reflection

TAAA027SAS

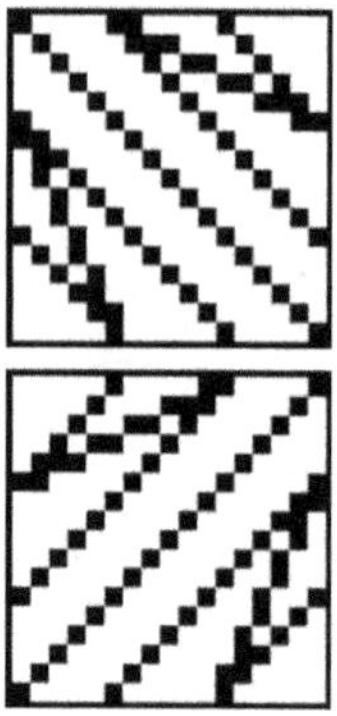

TAAA027 – Horizontal Reflection

TAAA027ASS

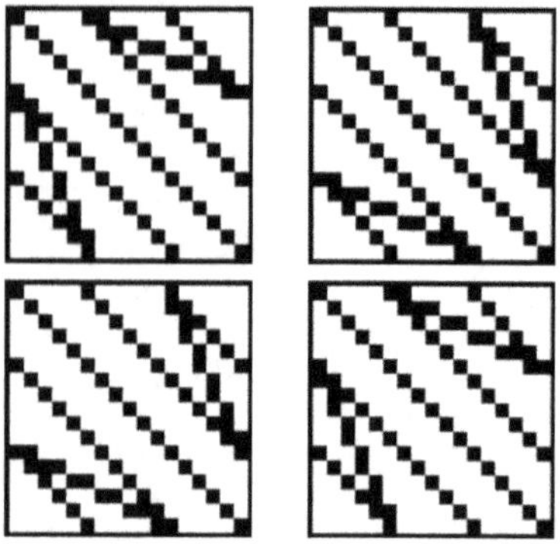

TAAA027 – Rotation and Vertical Reflection

TAAA027SAA

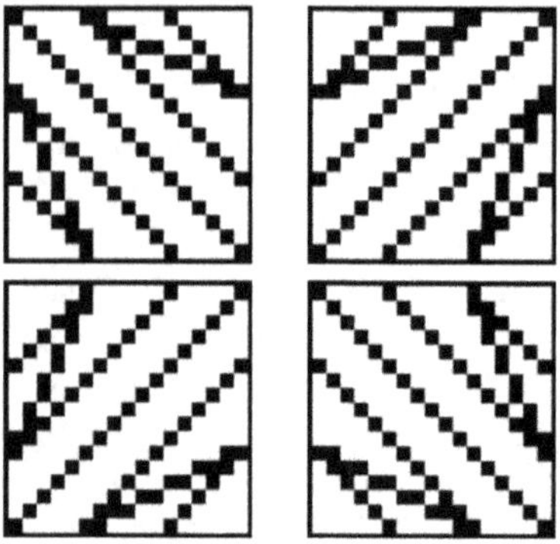

TAAA027 – Vertical and Horizontal Reflections

TAAA027AAS

www.ingramcontent.com/pod-product-compliance
Lightning Source LLC
Chambersburg PA
CBHW072313240926
51846CB00011B/133
* 9 7 8 1 9 6 0 7 4 1 0 1 1 *